FIRE and ICE

A History of Comets in Art

The appearance of comets is too beautiful
for you to consider an accident. . . .
(Seneca, *Natural Questions*, VII)

A Blazing Star,
Threatens the World with Famine, Plague and War;
To Princes, death; to Kingdoms many crosses;
To all Estates, inevitable Losses;
To Herdmen, Rot; to Ploughmen, hapless Seasons;
To Sailors, Storms; to Cities, civil Treasons.
(Du Bartas, *La Semaine*)

Frontispiece. John Martin, detail of *The Eve of the Deluge*, 1840, oil. The Collection of H. M. The Queen. (See Fig. 79.)

FIRE and ICE

A History of Comets in Art

ROBERTA J. M. OLSON

Published for the National Air and Space Museum
Smithsonian Institution
by Walker and Company New York

Illustration credits are listed on page 129.

First published in the United States of America in 1985 by the Walker Publishing Company, Inc.

Book design by Sheila Lynch

Published simultaneously in Canada by John Wiley & Sons Canada, Limited, Rexdale, Ontario.

Library of Congress Cataloging in Publication Data

Olson, Roberta J. M.
Fire and ice.

Bibliography: p.
Includes index.
1. Comets in art. 2. Comets. I. National Air and Space Museum. II. Title.
N8217.C47047 1985 760'.04495236 85-7295
ISBN 0-8027-0855-2
ISBN 0-8027-7283-8 (pbk.)

Printed in Hong Kong by South China Printing Company.

10 9 8 7 6 5 4 3 2 1

Contents

Acknowledgments

The genesis of this book occurred in 1978 while I was teaching Giotto's Arena Chapel frescoes, when suddenly I noticed that Giotto's Star of Bethlehem was no simple star but instead a very brilliant comet. The artist's observational powers astounded me, as did my own blindness in not recognizing earlier this apparition for what it was. The oversight results from two factors: (1) the quality of great art to slowly and continually reveal new insights each time it is observed, and (2) the lack of art historical/cultural studies on comet imagery and hence even an awareness of them.

The decision to research Giotto's comet was instantaneous. During the process, I was incredulous that no other person had commented on Giotto's star/comet. My findings were subsequently published in 1979 in *Scientific American*. I am most grateful for the public response to that study and for the European Space Agency's decision to name their satellite, designed to rendezvous with Halley's Comet in 1986, *Giotto*. After 1979, research on the general topic of comets continued. This book and exhibition include a representative selection from the material.

Walter J. Boyne, Director, and Mary Valdivia, Curator of Art at the National Air and Space Museum of the Smithsonian Institution, deserve a very special thanks for their enthusiastic support of both this book and the related NASM exhibition of the same title, which I am guest curating.

I should also like to thank the National Endowment for the Humanities for their generous Fellowship during 1982–83, which enabled me to gather material for this book and to begin giving it form. Several grants from the Committee on Faculty Research and the Mellon Committee of Wheaton College allowed me to not only pursue early stages of the research but also to continue on the project over the years and to purchase black and white photographs. There is also a lengthy list of private collectors and institutions, museums and libraries, whose valuable collections aided me immeasurably. A special note of thanks must go, however, to all the personnel of the New York Public Library and the Watson Library of the Metropolitan Museum of Art, especially to Pat Coman and William Walker.

There are so many helpful individuals who gave graciously of their time and expertise in every aspect of this book, that it is impossible to cite all their names. But my gratitude is nonetheless felt. Alexander B. V. Johnson, who holds the record for discovering the largest number of comets in works of art, should be singled out, as should E. D. H. Johnson, another faithful comet-seeker, and Emma P. Hoops, an accomplished German translator. Other comet-hunters and contributors are: Robert

L. Patten, Elizabeth E. Roth, Joseph N. Marcus, Monawee Richards, Robert Rainwater, Janet Backhouse, Adelaide Batchelor, Gisela Scheffler, D. J. Schove, John Ittman, Marzia Dall'Olmo, Frances Shirley, Mary Heuser, Travis Crosby, C. H. Beharrell, Ann Murray, Fred L. Whipple, Kerrie Chappelka, Alison Hart, Catherine Sieberling, Thomas J. McCormick, Ellen Hirschland, Jennifer Montagu, Ulrich Middeldorf, Colin Eisler, Konrad Kuchel, Laurie Vance Johnson, Charles Scribner III, Kauko Kahila, Mary Valdivia, Jane Shadel Spillman, John Rowlands, Carl Nix, Charles Thiels, Alice Frelinghuysen, Peter Volk, David J. Bryden, Patrick Moore, Rosalie Green, Richard A. Parker, Edith Porada, Peter Arms Wick, Marcello Violante, Rolf Biedermann, Marcy Sigler, Donald K. Yeomans, Margaret McCormick and Daniel Traister. Sotheby's in New York and London, as well as Christie's, New York, also facilitated my sleuthing.

A final note of thanks goes to: Richard K. Winslow for his thoughtful editorial suggestions, Julie Glass, Mary A. C. Fiske, Sheila Lynch for her attractive book design, and Timothy Seldes, for his belief in the project. Lastly, I am very grateful to Joan Silva, Nancy Shepardson and Kathie Francis for typing various sections of the manuscript as well as related material.

Title page illustration
from a treatise on comets
by Nicolaus Pruckner, 1532.

Foreword

In the long and outrageous history of human error and folly, the fear of comets holds a considerable space. Comets, right down to this very day, are held in terror by large numbers of people who consider them as harbingers of disaster.

To be sure, people are not foolish for no reason at all. Lying at the base of all superstitions and nonsense are observations and assumptions that seem to make sense—or even to be so obvious as to require no argument.

Let's try and work out the line of reasoning, for instance, that led to the fear of comets.

1) As far as our direct observation is concerned, the universe is a small place. Until quite recently in history, individuals knew only of a small section of Earth's surface in the vicinity of their dwelling places, and the sky did not seem very far overhead. The universe consisted then of a small patch of ground that was covered by a curved patch of sky.

2) Within this small universe, human beings seemed the most important material objects of all. It followed, therefore, that the universe must have been created to serve human beings and provide them with a comfortable home.

3) Changes took place in the sky that affected human beings. The coming of clouds presaged rain needed for crops, also violent storms that could do damage. The sun produced light and warmth and when it vanished from the sky, it was dark night and the temperature fell. The moon went through a cycle of phases that could be used to mark the passage of time and to guide human beings through the cycle of seasons—summer and winter, wet and dry, seed time and harvest.

4) The changes in the sky are cyclic and, after careful observation and thought, those changes could be predicted. The stars, generally, moved in regular circles and the sun and moon moved in predictable ways against those stars. In the early days of civilization, it was found that even certain bright stars (the planets), which moved in ways that seemed irregular at first, displayed a deeper regularity so that their movements could be predicted, too, though not as easily as those of the sun and the moon.

5) The simpler cyclic changes in the sky, those involving the sun and moon, seemed to match the obvious cyclic changes on Earth, such as the alternation of day and night, the changing lengths of day and night, the progression of the seasons, and so on. The subtler changes in planetary positions, compared with each other, with the sun and the moon, and with the back-

ground stars, must therefore indicate rather subtle changes in human affairs. Out of this reasoning was born the pseudo-science of astrology, which almost everyone considered as based on obvious truths, and which countless millions of people accept, even today, on the basis of this ancient reasoning.

6) There are unusual events in human life, however, that are not cyclic, but that are interruptions of the generally peaceful and expected progression. Usually, these are disasters—droughts, floods, epidemics, war, and so on. If everything human is predicted by the changes in the sky, these unpredictable disasters should correspond to something unpredictable in the sky as well.

7) The most obvious unpredictabilities in the sky are represented by the comets. They come without warning; they travel across the sky in an unpredictable path; they disappear at an unpredictable time. What's more, they have a shape that is different from all other heavenly objects, and that shape is itself portentous. A comet with its long sweeping tail looks like the head of a woman with long, streaming hair (a traditional sign of mourning), or like a sword (a sign of war and death).

8) The conclusion of all this, then, is that comets appear in the sky in order to serve as a warning of disaster. They are sent perhaps by the gods as a kindly gesture, to give humans time to change and stop sinning so that the divine anger would be turned aside and the disaster averted.

9) The final test of all this would be to observe whether disaster did, in fact, follow the appearance of a comet—and it always did.

With time, however, new facts arose. The nature and behavior of comets came to be understood. In particular, the orbits of many were worked out and it was clearly shown that their comings and goings were as regular as those of other heavenly objects. They were *not* unpredictable.

Secondly, the relationship between the planetary positions and human affairs (other than the obvious ones involving the sun's light and warmth and the moon's phases and tidal pull) became less and less likely as scientists understood more and more about the properties and motions of the planets. The assumptions on which astrology is based are simply false, and astrology is dismissed as nonsense by astronomers.

And what about the fact that disaster always follows a cometary appearance?—Well, disasters invariably come even when visible comets are *not* present in the sky. The sad fact is that disasters come *every* year and comets have nothing to do with it.

To fear comets *now*, therefore, is simply foolish, and yet—people do.

Roberta Olson, in this book, supplies us with a diverting account of people's fascination with and dread of comets, and how, as a result, comets have not only contributed to folly, but have filled human art and literature, to the enrichment of both.

Isaac Asimov

Fig. 1. Claude Nicollier, *Comet Bennett*, photographed at Gornergrat Observatory, Switzerland, March 26, 1970.

Introduction: On the Nature of Blazing Stars

Comets are nature's icons (Fig. 1)—single, wondrous images of awe and reverence that invariably inspire a sense of veneration and, often, of fear as well. Dramatically tossed up to illuminate the dome of the night sky, these "blazing stars," as they have long been called, violate what is perceived to be the predictable, elegant order of the heavens. Since the beginning of recorded time, they have fascinated all those who have seen them.

Unlike other spectacular celestial phenomena, such as eclipses, some comets are visible for months on end, and their fiery tails can stretch across vast reaches of the sky. It is no wonder then that many observers throughout the centuries have become obsessed with the awesome beauty of comets and have attributed deep significance to their appearance. Comets thus have an enduring presence in Western art and literature, as their images serve as portents, omens, and symbols; and so it is that people's myths and cultural attitudes toward comets can best be traced in the visual arts.

Astronomers find that the immediate source of comets is a vast cloud of them known as the *Oort Cloud*, which is gravitationally part of the solar system and extends thousands of times farther from the sun than the outermost planets. This cloud represents the remains of the primordial matter from which the sun and planets were formed some 4.6 billion years ago. In 1950, Fred L. Whipple, a leading American astronomer, deduced that the heart of a comet, its nucleus, is a "dirty snowball," typically a few miles in diameter. His theory has been amply verified by radar and modern observations of comets. A gravitational disturbance, such as the mass of a passing star, can occasionally dislodge one of these "snowballs" from the Oort Cloud and send it gliding toward the sun to become entrapped in a new orbit.

Most of the comets that have been observed from the earth, including Halley's Comet, travel in enormously elongated elliptical orbits that usually lie in planes that are highly tilted with respect to the orbits of Earth and the other planets (Fig. 2). The comets that round the sun in periods of two hundred years or less are called *short-period* comets. Astronomers find that of the six hundred or so comets with identified orbits, only about one hundred short-period comets are known. *Long-period* comets are those with periods over two hundred years; five hundred belong to this category, many of which have been calculated to take hundreds of thousands of years for a round trip. Some comets become nonperiodic, moving in open-ended paths in the shape of hyperbolas. After rounding the sun, they glide off into interstellar space—never to return.

The point in a comet's orbit closest to the sun is termed the *perihelion* (*peri* meaning "near," *helios* meaning "sun"), while the farthest point from the sun is called the *aphelion*. A comet's speed increases as it approaches the sun. A comet might race past its perihelion at a million miles an hour and creep along the most distant portion of its orbit at the speed of a bicycle. The orbit of a comet can be disturbed by the gravitational pull of the planets, altering its

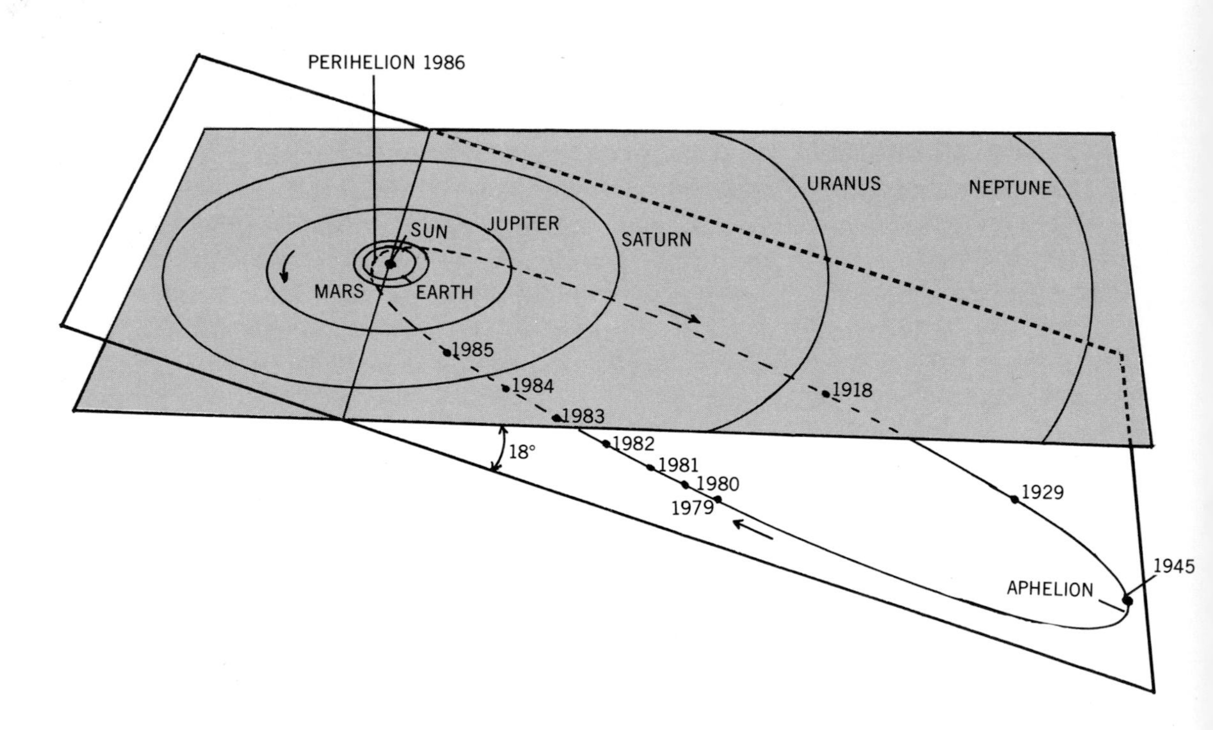

Fig. 2. Orbit of Halley's Comet.

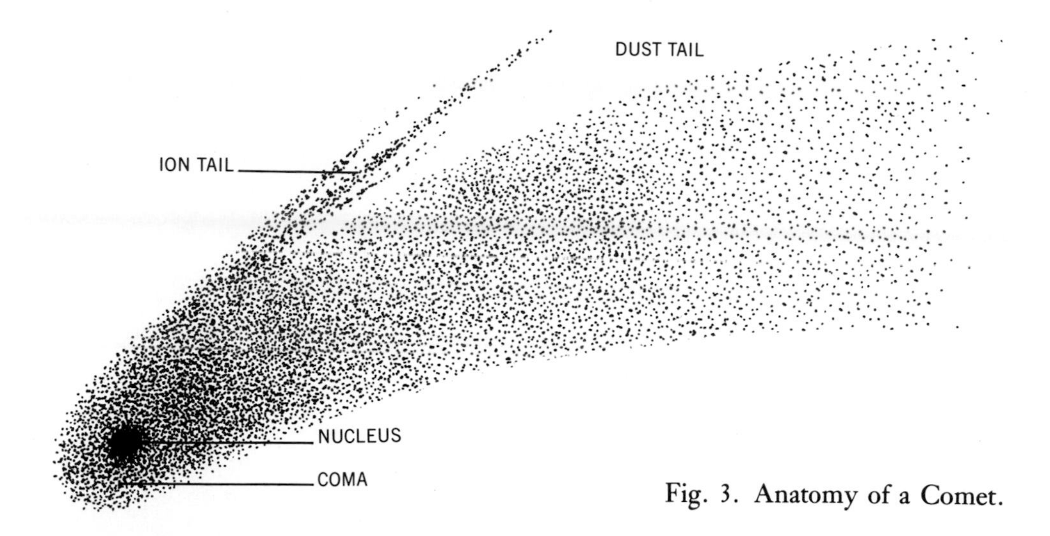

Fig. 3. Anatomy of a Comet.

course, speed, and brilliancy. On occasion, such a disturbance, or perturbation, can cast a comet off into space or send it plunging into the sun. During each passage around the sun, comets tend to lose perhaps a thousandth or more of their mass. Some comets, such as Biela's Comet in 1846, freakishly break up into fragments and disappear forever. A few, like the Great September Comet of 1882, are *sungrazers,* a term for comets that pass near the solar surface.

Comets are anatomically comprised of three major parts (Fig. 3). The first is the icy *nucleus,* usually a few miles in diameter. As the nucleus comes near the sun and encounters its heat, some of its frozen gases evaporate, releasing solid particles, and creating a surrounding halo of diffuse material thousands of times the size of the nucleus. This halo, the second main part, is called the *coma.* The coma and the nucleus together are referred to as the *head.* The fuzzy coma both reflects and fluoresces sunlight. It may be as large as that of the Great Comet of 1811, which had an estimated diameter of 1,250,000 miles. The third section, the *tail,* is the most flashy and is responsible for the name *comet.* The word is derived from the Greek word *komētēs* which literally means "long haired," a reference to the appearance of its manelike tail. As the comet comes within the planetary orbits, solar radiation and the outward flow of electrically charged particles, called the *solar wind,* sweep back the material of the coma, forming the magnificent tail. The tail always flows away from the sun.

There are two types of tails, and comets frequently exhibit both kinds simultaneously (Fig. 3). The *dust tail* forms a slightly curved, often billowing, tail which reflects the light of the sun. Many comets, especially the brightest, display another tail, the *ion* or *plasma tail,* which is virtually straight and is comprised of positively charged gas forced away by the solar wind. It streams away at a faster rate and thus tends to be slimmer than the dust tail.

Fig. 4. *Meteor Seen at Paddington*, 1850, mezzotint. Because of the explosive head, this object over London may more properly be identified as a bolide.

Like the gas of the coma, these ion tails shine by fluorescence. A comet may have several ion tails as did Donati's Comet in 1858. Comets are truly, as described in medieval times, "torches of the sun."

Every comet is unique, and its appearance fluctuates during each apparition. The appearance depends on several things: the size and shape of the nucleus, its orbit, the sun's effect on the escaping material, and the angle and distance from which we see it, etc. Tails also vary in length and character during each apparition. They may extend as far as the tail of the Great Comet of 1843, which had a two-hundred-million mile tail—the longest ever recorded. Tails can also split into parts with fanning sections, sometimes as many as twelve in number. Some comets (usually the smaller ones) never develop a tail but, rather, look like fuzzy patches of luminescence in the sky.

Over the ages, a variety of astronomical events, loosely referred to as "blazing stars," have been confused with comets by artists, writers, and humanity in general. These include meteors (Fig. 4), fireballs (particularly vivid meteors), bolides (meteors that seem to explode), and the aurora borealis. If an unusually bright light appeared in the heavens, with or without a tail, it was often labeled a comet. Previous to those of the eighteenth century, Western records on comets are not entirely reliable. Chinese astronomical records, on the other hand, tend to be more accurate and have been consulted usefully for specifics on earlier apparitions. In the eighteenth century,

Jesuit missionaries brought these annals to the West, and they were first published in 1846.

Meteors are, of course, very different in nature from comets. They are small bits of interplanetary matter, much of which may be comet debris cast off from the nuclei of comets, traveling around the sun. When they enter the earth's atmosphere, at speeds as high as forty miles per second, they burn up, appearing as streaks of light. Each meteor can be as minute as a grain of sand, yet its bright track is visible one hundred and fifty miles away. In medieval times, these missiles from space were given such fanciful names as "flying dragons," "serpents," and "heavenly flames" and, along with comets, were often depicted as such.

Comets lose material each time they pass the sun, leaving a wake of comet litter, some of which remains in the comet's orbit. When the earth passes through the orbit of a living comet, as it does twice a year in October and

Fig. 5. Albrecht Dürer, *The Opening of the Fifth and Sixth Seals*, 1511, woodcut.

Fig. 6. Giorgio de Chirico, *The Apocalypse*, n. 9, 1941, lithograph.

May in the case of Halley's orbit, we see meteor showers. These Halley showers are known respectively as the Orionids and Eta Aquarids. The Leonid meteor showers, whose orbit the earth passes through in mid-November, were reported to be so bright on the night of November 12, 1833, that people were unable to sleep and thought the sun had risen. Again, in 1966, the unpredictable Leonids staged an impressive show over the United States when over one hundred thousand meteors per hour were observed in Arizona. This phenomenon truly merits the medieval nomenclature of "flying angels" or "run of stars." Two other meteor showers are the Bielids, which are associated with the breakup of Biela's Comet in 1846, and the Perseids, the most famous of all the showers, displaying from five thousand to twenty-five thousand meteors per hour under optimum conditions.

Extraordinarily unnatural and frightening displays of meteor showers were used by artists throughout the ages as a visual device to depict a passage from John's description of the Apocalypse in Revelation (6:12–13): "And I beheld when he had opened the sixth seal, and lo, there was a great earthquake; and the sun became black as sack cloth of hair, and the moon became as blood; And the stars of heaven fell onto the earth. . . ." Two diverse works of art illustrating this image are a sixteenth-century woodcut by Albrecht Dürer (Fig. 5) and a twentieth-century lithograph by the surrealist Giorgio de Chirico (Fig. 6).

Large meteors that survive their fall through the earth's atmosphere and hit the ground are known as *meteorites*. A large one created a crater of about one mile in diameter, which is still visible today near Winslow, Arizona. Much larger meteor craters have been identified.

It is theorized that, some sixty-five million years ago, all the celestial bodies in our solar system were bombarded by comets and meteors. Evidence of this bombardment is found in the pockmarked appearance of the moon and planets. One current theory holds that a comet or an asteroid collided with the earth in that era, creating a huge cloud that surrounded our planet for several years and led to the extinction of the dinosaurs, and many other species. The deep gloom drastically reduced photosynthesis, destroying the plants upon which the giant herbivores fed, and the food chains supporting the dinosaurs and many other species collapsed. Paleontologists have found evidence of this pervasive dust cloud in a residual mud layer lying between strata of sedimentary rock dating to that period.

Early in this century, on the night of June 30, 1908, the mysterious Tunguska Event occurred in a remote region of Siberia. Since the Russian Revolution and World War I prevented investigatory teams from penetrating the area of the Tungus Forest until 1927, there is little contemporary evidence of what exactly happened. What is known is this: Something exploded, leaving trees within many square miles flattened, knocking people over more than thirty miles away, and even halting, for a time, trains on the Trans-Siberian Railroad. Since there are neither traces of radioactivity, nor a crater, nor me-

Fig. 7. *Encounter Between the Earth and a Comet of 13 June 1857*, from "Actualités Astrologiques."

teoritic fragments, some scientists believe that this blast, thought to have been the equivalent in energy to a small nuclear explosion, might have been caused by the explosion of a colliding comet, perhaps from the periodic Comet Encke.

Today, this conclusion is supported by reports from that time of a very bright fireball which was observed before the explosion was heard over six hundred miles away. Flames and smoke were seen over the site, incandescent matter was thrown to heights of over twelve and a half miles, and the nights after the explosion were unusually bright in western Asia and Europe. From the pattern of damage, it seems that the explosion would have occurred over five miles above the ground, and that the comet nucleus would have been about 130 feet in diameter. The Tunguska Event is an instance of a catastrophe commonly feared the world over and given visual form in the nineteenth century (Fig. 7). On a popular level, the idea of a collision of a comet with the earth is frequently linked to the most famous comet of all, Halley's Comet.

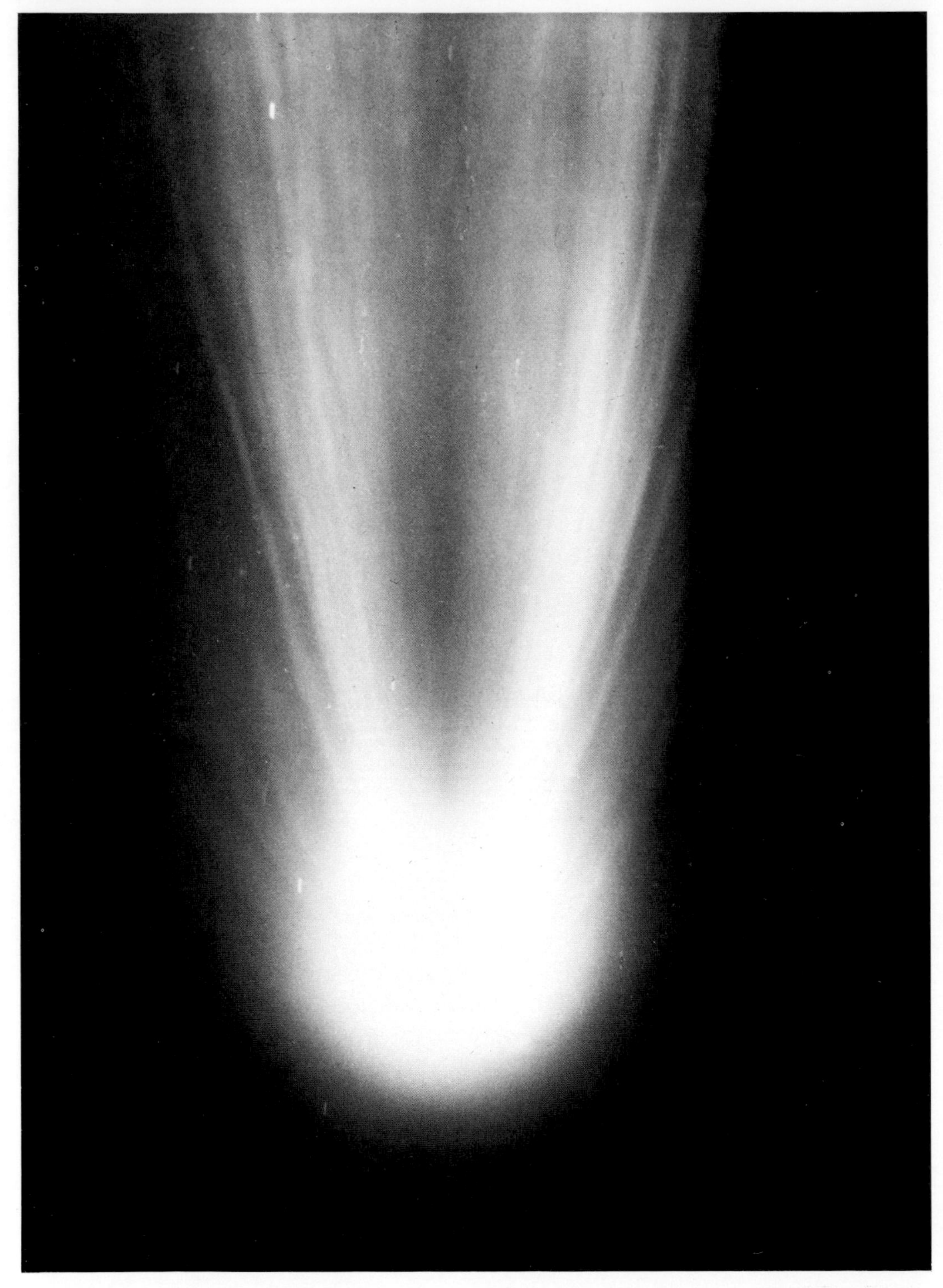

Fig. 8. *Halley's Comet, May 8, 1910*, photographed with the sixty-inch reflecting telescope on Mount Wilson.

One

One in a Hundred Billion: Halley's Comet

Of all the comets in the sky,
There's none like comet Halley
We see it with the naked eye and periodically.

—Anonymous

One hundred billion comets are thought to populate our solar system, although a mere seven hundred have been discovered to date. The most celebrated comet in history, Halley's Comet (Fig.8), is surrounded by a magical aura of tradition and fascination. Halley's (rhymes with alleys) is readily visible to the naked eye, and its appearance can be stunningly spectacular. Its returns are dependably regular. In fact, the comet we now identify as Halley's was the first to be recognized as periodic, returning to the vicinity of Earth approximately every seventy-six years.

Quite simply, timing is a major component of its mystique. Its regularity is virtually constant, except for a tantalizing variance of two and a half years, a result of slight changes in its orbit caused by the gravitational pull of Saturn and Jupiter. Its period is, therefore, just about that of the average human life span—a little longer than the biblical three score and ten—evoking both our own mortality and the inexpressible wonders of the universe.

People anticipate its arrival with great excitement. Mark Twain was born in 1835, a year of Halley's appearance, and he wrote humorously about a comet in *A Connecticut Yankee in King Arthur's Court*. He wanted desperately to live until the next appearance of Halley's Comet, and then to die with it—or so he dramatically claimed after suffering a series of heart attacks. His wish was fulfilled in a neat circle, for his life ebbed on April 21, 1910, one day after Halley's Comet reached perihelion. As Mark Twain himself put it, "The Almighty has said, no doubt, 'Now here are these two unaccountable freaks;

they came in together. They must go out together.' Oh and I am looking forward to that."

Customarily, a comet is named after the first person to spot it in the heavens. Why, then, was this most famous comet christened after the British astronomer Edmund Halley (Fig. 9)? Halley (1656–1742) had the good fortune to live at a time when comet studies had reached a crucial stage; his discoveries were part of the wide-ranging scientific breakthroughs of the seventeenth century. Halley's interest in the subject was sparked by the comet, later known as Halley's, that he tracked from his private observatory in Islington, near London, in 1682 (Fig. 10).

His theories about comets took many years to formulate and involved initially a rather complex collaborative relationship with Sir Isaac Newton (Fig. 11). By the 1680s, Newton had formulated his laws of motion and gravitation, which explained, among other things, the paths of bodies around the sun. Newton had based much of his gravitational theory on the orbit of the Great Comet of 1680, which also piqued Halley's interest. Halley and Newton corrected Johannes Kepler's theory that comets travel in straight lines, and they formulated hypotheses concerning the various orbits that comets could travel. The time was ripe for a serious definition of cometary orbits, for other astronomers had also suggested that comets move in an elliptical manner, but had made no attempt to apply this idea, in good scientific practice, to the observed data.

Their collaboration went far beyond the study of comets. Halley, who was

Fig. 9. Richard Phillips, *Portrait of Sir Edmund Halley* ca. 1721, oil. London, National Portrait Gallery.

Fig. 10. *Halley's Comet of 1682 Over Augsburg*, engraving.

a man of action and lofty principles but not of means, generously paid for the 1686 publication of Newton's monumental work, the *Principia*, which Newton had the means but not the motivation to publish. Halley played a major role in the preparation of Newton's manuscript for publication, and in his dedicatory ode (originally in Latin) wrote:

> . . . Now we know
> the sharply veering ways of comets, once
> A source of dread, no longer do we quail
> Beneath the appearance of bearded stars.

In the *Principia*, Newton also ventured to describe the physical nature of comets, concluding that the tail was formed by a flow of minute particles that emanated from the head under the influence of solar rays—an explanation that agrees substantially with modern theories.

Halley—an explorer, naturalist, and inventor as well as an astronomer—put aside his comet calculations until 1703, when he began to question whether the comet he had seen in 1682 had been observed at other times. After examining historical records, and making extremely difficult orbital calculations, he concluded that the comets seen in the years 1607 and 1531, together with the comet recorded in 1456, were the same celestial object as the comet he saw in 1682. He even accounted for the different intervals between perihelia by calculating the disruptions of its orbit, as well as its retrograde movement. Halley demonstrated once and for all that comets are members of our solar system that travel in eccentric elliptical orbits. In 1705 the Royal Society published Halley's original and profound findings as the *Astronomiae Cometicae Synopsis*, which he concluded by stating, "I may, therefore, with confidence predict its return in the year 1758. If this prediction is fulfilled, there is no reason to doubt that other comets will return."

Halley did not live to witness the fulfillment of his prophecy in 1758, a year in which, according to the French writer Voltaire, no astronomer slept. They were all too busy looking to the heavens for visual confirmation of Newton's theory and Halley's prediction. On Christmas night of that year, an amateur German astronomer, Johann Palitzsch, discovered the comet with his homemade seven-foot telescope. Halley's name was quickly attached, and astronomers throughout the world subjected it to intense scrutiny as it completed its swing around the sun in early 1759.

The first confirmed record of this comet's apparition dates to 87 B.C., when Julius Caesar was thirteen years old. Most cometologists, however, believe that Halley's Comet may have been sighted as early as 240 B.C., and new evidence in a Latvian folk song, confirmed by computer calculations, supports this view. A stanza of the song relates that "the sun lashed the moon with a silver broom." Other astronomers push the earliest known apparition back to 467–466 B.C., to a Chinese recording of an unspecified celestial phe-

Fig. 11. *Wedgwood Plaque of Sir Isaac Newton*, 18th c. Birmingham, City Museums and Art Gallery. The comet may be the Great Comet of 1680, upon which Newton based much of his gravitational theory, or Halley's Comet of 1758.

Fig. 12. *Halley's Comet of 1066*, detail of the Bayeux Tapestry, 1073–83, 20 in. high. Bayeux, Town Hall. The inscription reads: *They are in awe of the star*. With special authorization from the city of Bayeux.

nomenon, while yet others would like to confirm even earlier sightings. With the aid of modern technology and the help of early Chinese records, all of the comet's returns to perihelion have been firmly calculated back to 240–239 B.C. as listed. (Because of the changes in the calendar and methods of computing years B.C., the first four apparitions may be listed as 239, 163, 86, and 11 B.C.)

240 B.C.	A.D. 530	A.D. 1301
164	607	1378
87	684	1456
12	760	1531
A.D. 66	837	1607
141	912	1682
218	989	1759
295	1066	1835
374	1145	1910
451	1222	

The first useful Western records are Roman reports of Halley's Comet in 12–11 B.C. (Cassius Dio Cocceianus, *Roman History*). In A.D. 66 the comet reappeared while Jerusalem was under siege by the Romans. The Jewish

Cometa

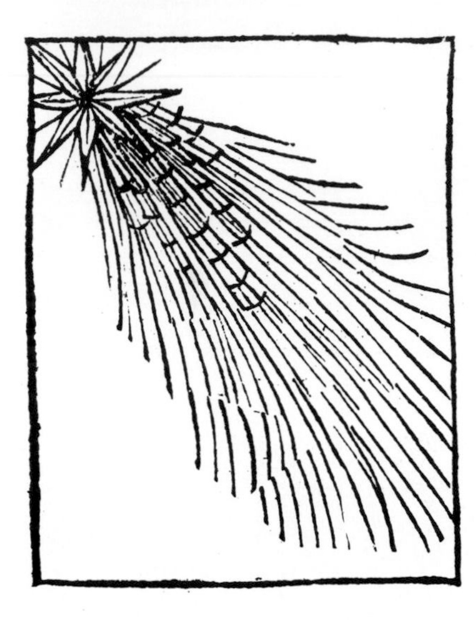

Fig. 13. *Halley's Comet of* A.D.684, from Hartmann Schedel, *The Nuremberg Chronicles*, 1493.

historian Josephus described it as a sword that hung over Jerusalem for a whole year, prophesying the destruction of the city four years later. According to past observations, Halley's Comet, while easily predictable, is not always brilliant, because of the earth's varying positions when it makes its return. The comet made a relatively poor showing in 1378, for example, although it was spectacular during its 1301 and 1456 apparitions.

The 684 apparition is the first known to have been described in visual form (Fig. 13). But there's a catch; while the woodcut purports to reproduce the comet in 684, it was designed eight centuries after the fact and published in 1493 in the *Nuremberg Chronicles (Liber Chronicarum)*, a history of the world and the city of Nuremberg. The stylized bold print, or a close variant of it, also reappears throughout this volume to illustrate other comet apparitions.

The earliest depiction of Halley's Comet executed at the time of the apparition—and perhaps the most famous one—occurs in the Bayeux Tapestry (Fig. 12), portraying the 1066 apparition. The "tapestry" is actually crewel embroidery on eight narrow strips of coarse linen, approximately 231 feet long by 20 inches high, which was executed between 1073 and 1083. It was reputedly commissioned by Queen Matilda, wife of William the Conqueror, to illustrate her husband's victory at the Battle of Hastings in 1066. Numerous contemporary accounts comment on the sensational nature of the comet, which looks like a primitive rocket spewing forth flames. They describe it in such rhetorical phrases as "a torch of the sun" or "a flaming beam," language that complements the Romanesque stylizations of the Bayeux Tapestry. These flattened and simplified cartoonlike forms endow the image with a symbolic power worthy of the accompanying inscription: *They are in awe of the star*. Indeed, the 1066 passage of the comet inflicted mass terror over all of Europe. It was said to portend all manner of calamities, including war and the death of kings. Conveniently, King Harold II perished at the Battle of Hastings. But the comet was a double-edged sword, for it also augured victory for the Normans, which changed the face of England forever.

After the Bayeux Tapestry, visual records of all but two apparitions (those of 1378 and 1607) are preserved. In the next century, the 1145 appearance is recorded in a spartan line drawing from the *Eadwine (Canterbury) Psalter* (Fig. 14), an English manuscript collection of the Psalms. The drawing's position and scale suggest that this charming image was added after the text, as does the Old English inscription, which appears to have no connection with the text. Recently, it has been suggested that the 1222 apparition of Halley's Comet is preserved in a newly discovered fresco fragment (c. 1250) in the Palazzo della Ragione, Mantua, signed by an otherwise unknown Byzantine artist, Grixopolus.

Without question, the most remarkable early observation of Halley's Comet was painted by Giotto, the Florentine pioneer of naturalism (Figs. 15 and 16). He immortalized the 1301 apparition in his depiction of the Adoration of the Magi in his Scrovegni Chapel series in Padua (1303–06). Since this unprecedented representation ranks as the first visually convincing representation of a particular comet, it can in a way be thought of as a "portrait." Like people, comets have unique, recognizable characteristics and profiles; there is no doubt that Giotto recorded the visage of Halley's Comet in his fresco. In addition, the artist's brushstrokes have effectively captured the pulsing energy of the comet's coma, as well as its center of condensation, which is frequently seen within the more diffuse coma.

Giotto painted the comet in place of the more conventional Star of Bethlehem. He thus married his naturalistic impulses with an arcane tradition

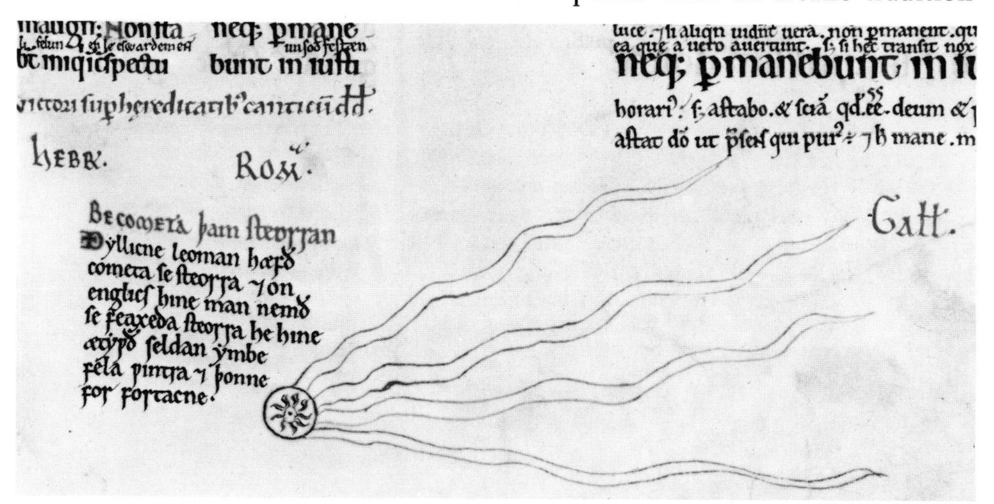

Fig. 14. Eadwine, *The 1145 Apparition of Halley's Comet*, *The Eadwine (Canterbury) Psalter*, Fol. 10r. Cambridge, Trinity College Library. The legend on the left refers to the radiance of the "hairy star" and reports that comets appear rarely and, when they do, as portents.

Fig. 15. Giotto, *The Adoration of the Magi*, 1303–06, fresco, 78¾ x 72¾ in. Padua, The Scrovegni Chapel. The comet in Giotto's work is Halley's Comet of 1301. Interestingly, the artist first depicted the bright center of the comet's head as a more conventional eight-pointed star and then built up layers of pigment on top, to create the illusion of translucence, as in nature.

Fig. 16. Giotto, *Halley's Comet of 1301*, detail of Fig. 15. This close-up reveals the manner in which the artist applied tempera and gold pigments to the plaster wall in textured strokes to approximate the visual appearance of the coma and tail.

dating back to the time of the Church Fathers Origen and John of Damascus that claimed the Star of Bethlehem was in fact a comet. This idea was also preserved through popular medieval writers such as Jacobus de Voragine in his *Golden Legend*, and probably depended upon the pagan belief that a comet apparition signifies the birth of a king. Giotto's magnificently vivid comet, which dominates the classical fresco, helped to evoke for fourteenth-century viewers the strong emotions and significance of the scene; for they, like the Magi, had recently witnessed an awesome celestial event, the unforgettable apparition of Halley's Comet in 1301. The eminent historian of the time, Giovanni Villani, wrote that the *stella comata* appeared in the heavens in September "with great trails of fumes behind" and remained visible until January 1302. While there are discrepancies about the length of time the comet was visible, there is virtual agreement that the comet spread across more than a third of the sky.

Padua, a university town where Galileo would eventually hold a chair, was renowned for the study of mathematics, a science that would help transform astrology into astronomy; and thus the town was a center for early astronomy. Certainly Giotto and/or his patron had access to some of the most recent theories about the heavens. Giotto's comet was such an advanced artistic statement that it was followed by less than a handful of tiny, timid imitations by his school. In subsequent centuries, the tradition of representing the Star of Bethlehem as a comet was only rarely revived, usually by artists with esoteric literary leanings.

(The above discussion does not mean to suggest that the Star of Bethlehem is now thought to have been a comet. Its nature is still debated today, and one plausible hypothesis is that the Star of Bethlehem was a brilliant conjunction of planets.)

Fig. 17. Diebold Schilling, *Halley's Comet of 1456*, *Lucerne Chronicles*, Fol. 61v, 1508–13. Lucerne, Zentralbibliothek. This folio shows the pernicious effects of the blazing comet.

Fig. 18. *Halley's Comet of 1531 in the Constellation Leo*, from Petrus Apianus (Apian), *Practica*, 1531.

The 1456 apparition of Halley's Comet (Fig. 17) sparked a great deal of impassioned reaction. Its appearance, according to popular legend, caused Pope Callixtus III to excommunicate it as an instrument of the devil. The pope did, however, urge people to pray three times a day and instituted special prayers to protect the devout. This comet was also blamed by Christians for the Turkish menace in eastern Christendom. These superstitious actions occurred during the supposedly enlightened age of the Renaissance, which witnessed the initial stirrings of modern astronomy well before the first telescope (invented 1609). The Florentine astronomer Paolo dal Pozzo Toscanelli, who influenced Leonardo da Vinci's writings on astronomy, optics, and perspective, traced in delicate line drawings in a manuscript now in the Biblioteca Nazionale, Florence, the progress of that terrible hairy star of 1456, against the backdrop of the constellations. It was also during this apparition that the Austrian scholar Georg Purbach diverged from the common practice of regarding the comet as an omen and attempted to calculate the distance of Halley's Comet from the earth.

The illustration of Halley's 1456 apparition from the *Lucerne Chronicles*, an amazing illustrated manuscript of civic history dating from ca. 1508–13 by Diebold Schilling, captures effectively some of the current fears associated with all comets in general (Fig. 17). The folio attributes to its baneful influence, in both verbal and pictorial form, monstrous births (two-headed animals, people with Down's syndrome), earthquakes, illnesses, and exotic red rain. Interestingly, the rather schematically rendered comet is depicted twice, as it was indeed visible both before and after perihelion.

Halley's Comet returned again in 1531, an apparition that was later to be one of the catalysts in Halley's theory of periodic comets. It is lionized in the stylized, albeit significant, illustration for a treatise by Petrus Apianus, known more commonly as Apian (Fig. 18). Later, in 1540, Apian published another tract in which he first postulated that the tail of a comet always points away from the sun. His significant idea is already portrayed here in 1531, where the comet and the sun are depicted in five exposures, as it were, in the constellation Leo. This idea is surprisingly accurate (although his triangular orbit is not) and contradicts in spirit the superstitious inscription below the illustration. Although the sixteenth century was an age torn asunder by devastating wars, superstitions, and the great doubts and upheavals of the Reformation, it was succeeded by the century that gave birth to modern science.

During the seventeenth century, the telescope made its debut in Europe. Longomontanus, who together with Johannes Kepler described the Comet of 1607 (later known as Halley's), noted its long, bright tail and boasted that it "equalled Jupiter in size and Saturn in brightness." The apparition of 1682, which heightened Halley's concentration on the subject of comets, was first spied by a German astronomer. In the seventeenth-century print that preserves this apparition (Fig. 10), a decidedly novel, more objective point of view can be detected. A group of people have gathered to gaze rather dis-

Fig. 19. Samuel Scott, *Halley's Comet of 1759 Over the Thames*, oil, 32¼ x 44 in. Private Collection.

passionately at the natural fireworks. They point to the comet with obvious interest and perhaps awe, but not with the customary panic.

Despite Halley's rational proof that comets are a part of the physical universe and travel in defined elliptical orbits, superstition continued to attend the 1759 apparition. A broadside published in Boston that year reveals that the ambivalent attitude about comets still prevailed; people could gaze at them with the latest scientific apparatus but nonetheless predict dire happenings. By the eighteenth century, sophisticated souls could regard comets more or less dispassionately, as demonstrated by an elegant mahogany drop-leaf table, in a private collection, with its legs cleverly decorated by a Connecticut craftsman with comet motifs commemorating the passing of Halley's Comet in 1759. The English painter Samuel Scott rendered a seemingly realistic vision of the 1759 apparition in a nocturnal scene (Fig. 19) where the comet hovers eerily over the Thames, with Westminster Cathedral in the background and the King's royal barge in the foreground. Even though the painting depicts a naturalistic landscape, the comet may serve as a traditional allusion to the beneficence of the English king. Ironically, the composer George Frederick Handel, who worked for King George I and whose *Water Music* was performed in England on the monarch's barge, died in London under the fiery aegis of this comet. Scott has rendered the comet's tail with a peculiar staccato configuration, which may reflect either the actual disturbances in the comet's tail, as reported by observers in 1910, or the artist's own stylization. This manner of depicting the tail, together with the comet's horizontal appearance, may be an English convention for rendering comets and meteors (see Fig. 65) and may, therefore, not record the comet as it actually appeared. Similarly, Scott, who established himself as a marine painter in oils and watercolor, only set foot in a boat once in his life!

When Halley's Comet returned in 1835, an interest in artistic realism was developing, reflected in the scientific drawings of the day. Sir John Herschel, the noted British astronomer, faithfully recorded its appearance in delicate, analytical drawings of its structure and envelopes, while F. W. Bessel made a series of scientific drawings that described the many volatile changes occurring in a comet's configuration. Despite the new objectivity, however, superstitions still persisted. An explosive broadside (Fig. 20) captures a more popular interpretation of the comet, echoed in Alfred, Lord Tennyson's *Harold*:

> Lo! there once more—this is the seventh night
> You grimly glaring, treble-brandished scourge of England. . . .
> It glares in heaven, it flares upon the Thames.
> The People are as thick as bees below,
> They hum like bees—they cannot speak for awe.
> Lord Leodwing dost thou believe that these
> Three rods of blood-red fire up yonder mean
> The doom of England and the wrath of Heaven.

Halley's Comet of 1835–36 was blamed for, among other things, the fire that all but destroyed New York City, the Zulu massacre of the Boers, and the Mexican massacre at the Alamo. The American Mennonites were convinced it foreshadowed the Apocalypse and hence allowed their crops to wither. The 1835 apparition may also have inspired comet patterns in both cut- and plate-glass, which were produced by several American companies during the height of comet fever around the middle of the century.

Photography finally captured its first image of Halley's Comet in 1910 (Figs. 8, 21, and 118). During the 1910 apparition, which coincided with the death of King Edward VII of England, astronomers at Yerkes Observatory in Williams Bay, Wisconsin, determined through spectral analysis that the comet's tail contained poisonous cyanogen gas. Because the earth was scheduled to pass through this "deadly" tail, people panicked. Some of the more ardent committed suicide, while a group in Oklahoma reputedly tried to sacrifice a virgin. Other, more sensible folk boarded up their houses or sealed their windows against noxious fumes. In various parts of the world, charlatans hawked "comet pills" and "comet insurance." (For the 1985–86 apparition, the Chaffee Planetarium in Grand Rapids, Michigan, is also selling comet pills, but with the following warning: "Musuem surgeon general has determined that worrying about comets can be hazardous to your health.") Talismanic medals were also issued in the time-honored manner. Various other wild rumors circulated: one held that the comet would strike the North Pole, alter the magnetic field, and electrocute everyone. The most intriguing notion was that during the comet's appearance, all the earth's nitrogen would be converted into nitrous oxide, otherwise known as laughing gas. The sybaritic set responded in kind: Bons vivants in Paris held "comet suppers," while The Comet Cocktail was served at the Plaza Hotel in New York City, a town in which were held many rooftop comet parties.

All comet lore aside, scientists approached the 1910 apparition with great fervor. Although A. C. D. Crommelin and Philip Crowell, two British astronomers, had made many preliminary calculations about the time and nature of its reappearance, the comet was first sighted early on September 12, 1909, by a German, Max Wolf, when it was still over three hundred million miles from the sun. During May, the comet did pass directly between the earth and the sun, apparently with no ill effects.

Fig. 20. *Broadside of the 1835 Apparition of Halley's Comet*, woodcut.

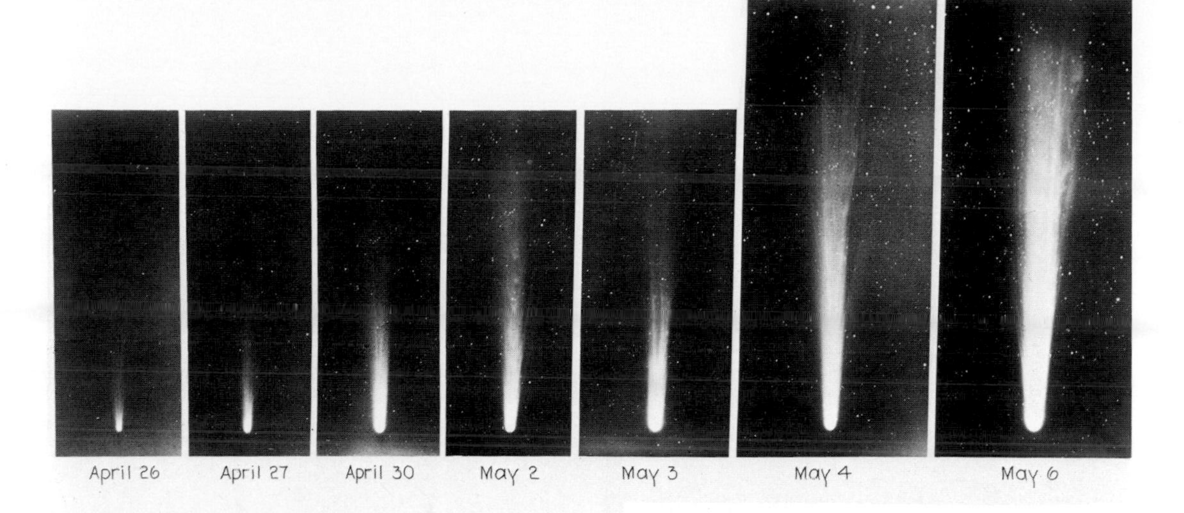

Fig. 21. *Fourteen Photographic Views of Halley's Comet,* made between April 26 and June 11, 1910. Perihelion was on April 19, 1910.

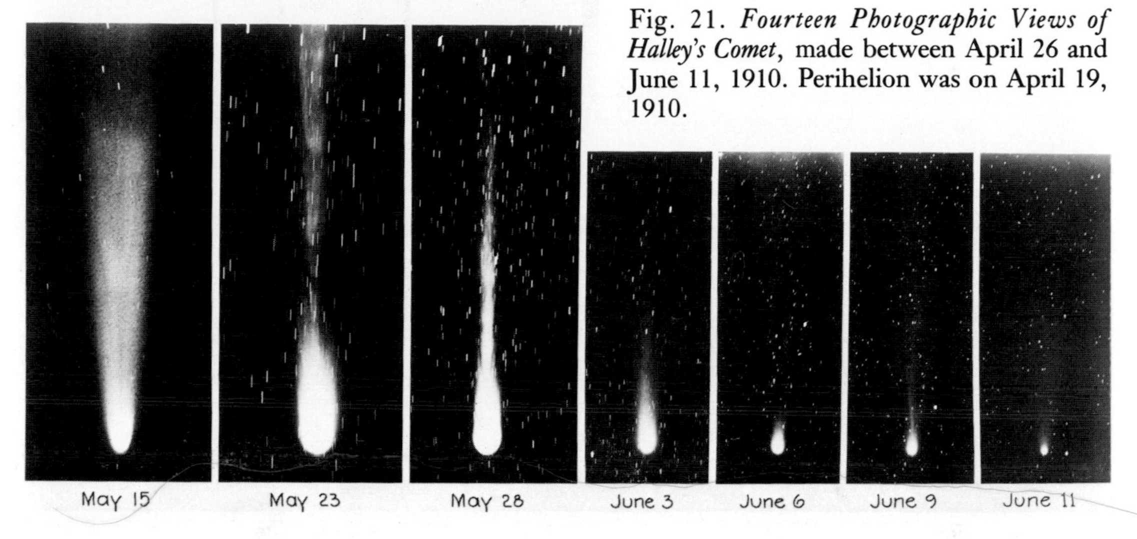

The 1910 apparition was not as spectacular as expected, however. Its impact was also perhaps lessened by comparison to the brilliant, nonperiodic daylight comet that had preceded it earlier that year. These two comets were later confused in the minds of some observers, while a Royal Delft commemorative plate preserves the two dazzling denizens of that year together.

Halley's twenty-ninth return since its first recorded apparition will be in 1985–86. Perihelion will occur February 9, 1986, when the comet will come within 55.8 million miles of the sun. Aphelion occurred in 1948, at the outermost reaches of its cigar-shaped orbit, 3.2 billion miles away. Because of the great strides in technology since its last swing around the sun, it was spotted extremely early during its swing back to Earth, on the night of October 16, 1982, by David C. Jewett and G. Edward Danielson of the California Institute of Technology. (The first amateur, Tsutomu Seki of Japan, found and photographed the comet on September 22, 1984.) An advanced electronic detector and a two-hundred-inch telescope at Palomar Observatory picked up the comet as a minute speck of light reflected off the vagabond chunk of dirty ice. At the time, it was still beyond the orbit of Saturn, eleven times farther from the sun than the earth. This early sighting has given astronomers an especially long time to study its habits.

Its return will witness international cooperation and compromise. The original, daring U.S. venture was unfortunately scrapped by governmental agencies due to a complicated set of circumstances. It would have been the most dramatic of all space missions, for it was to attempt an actual landing on the comet's nucleus. NASA has, however, reprogrammed a Pioneer space probe (which has been orbiting Venus) to observe the comet. In addition, an Explorer III satellite, rechristened *International Cometary Explorer*, has been reprogrammed to fly by the faint comet Giacobini-Zinner and to monitor the effects of the solar wind on it before Halley's Comet reaches perihelion. The United States is also participating in the Pathfinder program—the result of an agreement signed in fall of 1984 between NASA, Intercosmos (the U.S.S.R equivalent), and the European Space Agency—which will track the various probes being sent to the comet. Lastly, a NASA shuttle flight in March 1986 will observe the comet's hydrogen halo with three ultraviolet cameras during the encounters of the probes with Halley's Comet.

Three other groups plan probes to penetrate the mystery of Halley's Comet in March 1986, when it crosses the ecliptic plane one month after perihelion, traveling at about 21,600 MPH (Fig. 22). The first, sponsored by the European Space Agency (a consortium of eleven Western nations), has been named, for obvious reasons, *Giotto*. Its launch is planned for July 1985, and it will have eight different types of observing instruments on board. On March 13, 1986, the spacecraft will come within 310 miles of the comet's icy nucleus, which is only about four miles in diameter. Cameras will take detailed photographs of areas, some as small as thirty feet across. Japan plans to send up two spacecrafts, baptized *MS-TS* and *Planet A*, on January 4, 1985, and August 20, 1985, respectively. The Soviets, in partnership with the French, have launched *Vega 1* and *2* as part of the most ambitious mission to explore the comet. These are to rendezvous with the comet in early March and to approach within six thousand miles of the solid nucleus, where the plan is to spend several hours testing their instruments, including two television cameras. So stands the quest to conquer the comet as this book goes to press. Any number of alterations or revelations may occur between now and March 1986. All these probes will ultimately test the specifics of Whipple's generally accepted "dirty snowball" theory.

A ground watch has been organized, known as the International Halley Watch. It is comprised of a network of amateur and professional astronomers whose goal is to promote cooperation, communication, and standardization of information, and will be operated out of the National Aeronautics and Space Administrations's Jet Propulsion Laboratory in Pasadena, California, in conjunction with Dr. Remeis Observatory in the Federal Republic of Germany. Ray L. Newburn and Donald K. Yeomans are among the guiding lights and coordinating astronomers at Pasadena, and Jürgen Rahe is the di-

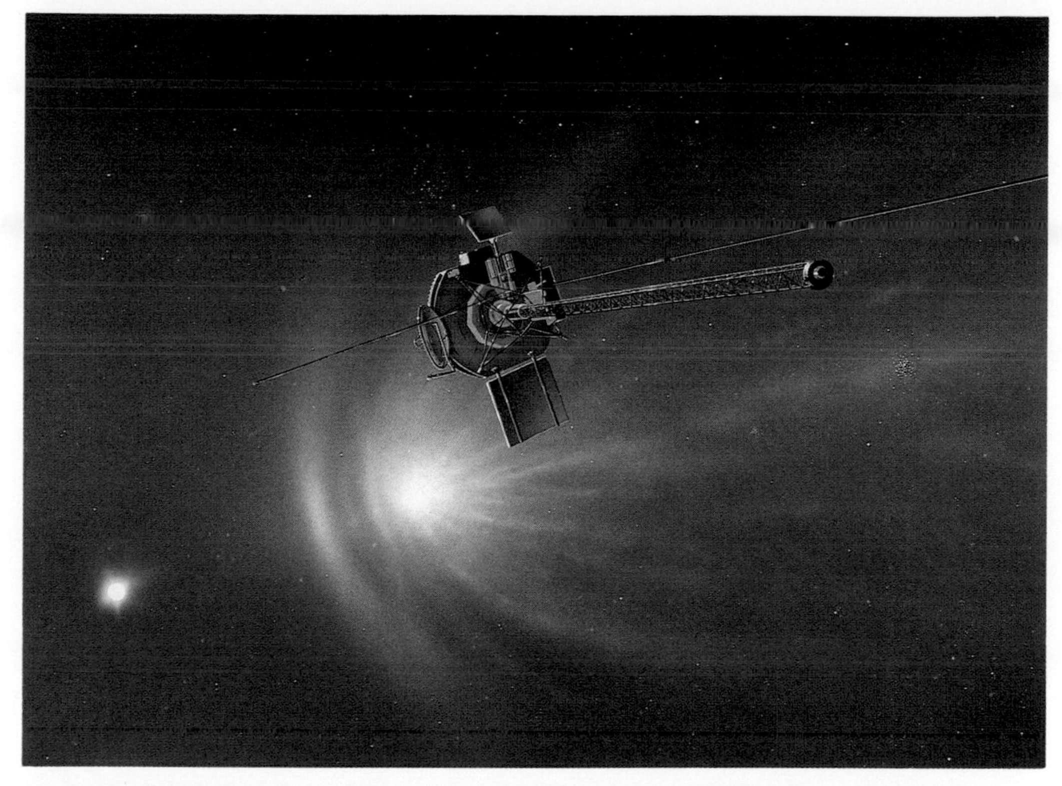

Fig. 22. Rudolf Brammer, *Rendezvous with Halley's Comet in 1986*, 1982, tempera on black paper, 19¾ x 27½ in. Stuttgart, Planetarium der Landeshauptstadt. This painting simulates the excitement surrounding the encounter between an earth satellite and the comet.

rector at Bamberg. According to recent calculations, the inbound comet will come within 57.7 million miles of Earth on November 27, 1985. The comet's closest outbound approach to Earth will be on April 11, 1986, at a distance of 39 million miles.

The 1985–86 apparition is expected to be disappointing as compared with previous ones, due to three factors: (1) the position of the earth in its orbit and the tilt of its axis; (2) pollution in the atmosphere; and (3) the vast increase in artificial lighting, especially near urban centers. But history is full of surprises. Scientists hope that technological advances will more than compensate for these factors. Since comets are thought to be among the most primitive bodies in the solar system, scientists will assuredly gain from the comet's visit a better understanding of the formation of our solar system some 4.6 billion years ago.

While Halley's Comet may not be spectacular in 1985–86, its return promises to be once again an unforgettable, "star-crossed" experience for those who witness it. After April 1986, Halley's Comet, the Old Faithful of the skies and an enduring symbol of wonder and beauty, will then go back to deep space, not to return again for another lifetime.

Two

Awesome Signs: From Antiquity Through the Renaissance

> No one is so completely slow and dull and stooping to the earth that he is not aroused by celestial phenomena. . . . Comets are nothing more but a pure fire which remains for six months at a time.
>
> —Seneca, *Natural Questions*, VII

In the mists of prehistory, people lived close to nature and were influenced by its unpredictable moods. Astronomy, the most ancient of the physical sciences, arose naturally from observations of the moon, planets, sun, and stars, and from the striking spectacles of eclipses and comets. Even the most primitive people must have found it desirable to take note of the cyclical movements of the heavenly bodies for such practical uses as planting and marking time and as guides for travelers. Ancient peoples—especially the Babylonians with their complex methods of sky-searching, the Egyptians with their stars and god-permeated heavens, the Greeks with their compulsion to explain the logical workings of things, and the Romans with their mania to control and conquer—were all fascinated by comets.

The Mesopotamians—the Babylonians, Assyrians, Chaldeans, and Sumerians—observed the skies constantly and developed a method for predicting lunar and planetary motions. Because their gods were astral deities, Mesopotamian astronomy was strongly tied to religion and controlled by the priestly caste, which developed it into a form of astrology. It was not an astrology of personal forecasts as we know it but, rather, a royal institution for the welfare of king and country. The priests foretold general events for large numbers of people, either by divination with sheep livers or by observation of the fixed stars, sun, moon, planets, and disturbances in the heavens. Comets and the most luminous shooting stars were considered important for their astrological effects and were observed and recorded on clay tablets.

Scientist Fred Whipple believes that Halley's Comet, near perihelion in late 163 B.C., was recorded on one of these tablets, which is now in the British Museum.

The ancient Egyptians, though not as observationally oriented as the Babylonians, may also have occasionally represented a comet. One image thought to be that of a comet has been found on the ceiling vault of a sarcophagus chamber, the Monument of Pedamenope, at Luxor. For Egyptians, astronomy was also controlled by a priestly class, and comets are mentioned in their hieratic papyri texts.

With the Greeks, whose astronomy was based more on philosophy than on observation, great strides were made with the application of new geometric techniques. Personal astrology and the systematization of the zodiac were also Greek contributions to astronomy. The Greeks had a variety of ideas about comets, but it was Plato's student Aristotle who prejudiced Western minds for thousands of years about them. In his *Meteorologica*, Aristotle wrote that comets were part of the atmosphere, dry exhalations of the ground that inhabit the upper air. He also began to catalog comets into descriptive types, encouraging others to do likewise even into the seventeenth century A.D.

Aristotle followed earlier Hellenic traditions for descriptions of comets. In the *Iliad* (XIX), Homer wrote, "The red star, that from its flaming hair shakes down diseases, pestilence and war. . . ." Aristotle echoed this attitude toward comets by stating that the appearance of a comet foreshadowed, among other calamities, a drought. The malevolently portentous nature of comets was therefore firmly established in Western thought until the Aristotelean spell was broken in the late sixteenth and early seventeenth centuries.

No Greek visual representations of comets are known. If any were ever produced, it is likely that they would have been painted in large-scale frescoes, none of which are extant save for the Macedonian murals recently discovered in the tomb of a ruler, perhaps Phillip of Macedon. Until the Late Classical Age, the Greeks were so preoccupied with the human body—especially that of the male—that they paid much less artistic attention to landscape until a relatively late date.

Among the most important Greco-Roman astronomers were those of the Alexandrian school, particularly Hipparchus (ca. 150 B.C.) and Ptolemy (ca. A.D. 140). Hipparchus, perhaps the greatest astronomer of antiquity, was inspired by a comet in 134 B.C. to form the first fixed-star catalog, which consisted of 1,080 stars. He also invented the astrolabe, the indispensable instrument used until the eighteenth century for plotting the longitude and latitude of celestial bodies. Ptolemy also compiled a star catalog, recorded a great many of Hipparchus's ideas, and wrote the *Almagest* (Arabic from a Greek word meaning "the greatest"), which was to become the Bible of astronomy until Copernicus changed people's view of the universe in the sixteenth century. Ironically, Ptolemy is remembered disparagingly for his

Fig. 23. Obverse and reverse of a coin commemorating Julius Caesar, ca. 30 B.C.–A.D. 14. Munich, Bayerisches Nationalmuseum, Basserman-Jordan Collection.

geocentric system, which tyrannized Western ideas about the solar system and universe for centuries. A pseudo-classical text, attributed to Ptolemy, lists nine types of comets according to their shapes. The Middle Ages and the Renaissance perpetuated this basic categorization of comets, sometimes expanding them into ten varieties.

The more pragmatic Romans did not add substantially to astronomy, but both Pliny the Elder and the enlightened Stoic philosopher Seneca commented extensively on all manner of comets. Seneca devoted an entire chapter of his *Natural Questions* to a dispassionate consideration of comets, arguing against the Aristotelean view and, in so doing, approaching a true understanding of their nature. He saw comets as celestial bodies and wisely foresaw that "men will someday be able to demonstrate in what regions comets have their paths, why their course is so removed from the other stars, what their size and construction. Let us be satisfied with what we have discovered, and leave a little truth for our descendants to find out."

But even the pragmatic Romans had a superstitious side, which perpetuated earlier traditions associating comets with wars, disasters (literally, "bad stars"), and plagues. In his *Natural History*, Pliny the Elder commented on the Comet of 48 B.C. that "we have in the war between Caesar and Pompey an example of the terrible effects which follow the apparition of a comet . . . that fearful star which overthrows the powers of the earth, showed its terrible locks." Kings' births, deaths, ascendencies, and falls could all be augured by comet apparitions, or comets could even be invented to accompany such political developments. Diodorus Siculus (*Library of History*, XVI) relates that "a torch blazing in the sky" singled out an individual as extraordinary. Both of these beliefs are echoed in a coin commemorating Julius Caesar (Fig. 23), whose reverse features an eight-pointed comet-star replete with a flaming tail and an inscription citing his posthumous deification, a necessary adjunct to imperial succession. As Suetonius wrote:

> He . . . was numbered among the gods, not only by a formal decree, but also in the conviction of the vulgar. For at the first of the games which Augustus gave in honor of his apotheosis, a comet shone for seven successive nights, rising about the eleventh hour, and was believed to be the soul of Caesar, who had been taken to heaven; and this is why a star is set upon the crown of his head in this statue. (*The Lives*, I-LXXXVIII)

This comet was the great daylight Comet of 44 B.C. At the death of subsequent Roman emperors, comets were also reported by Suetonius, Pliny, Tacitus, and Virgil, among other authors, and were regarded as torches to light the way to kings' tombs. Roman authors were not immune to satirical implications of the vulgar beliefs about comets, as seen in Juvenal's *Satires*, or in the Emperor Vespasian's remarks about the comet of A.D. 79: "This hairy star does not concern me, it menaces rather the king of the Parthians, for he is hairy and I am bald." Ironically, Vespasian died that very year and on his deathbed reportedly stated, "Woe's me. Me thinks I'm turning into a god" (Suetonius, *The Lives*, VIII–XXIII).

The mystery cult of Mithraism, one of early Christianity's rivals during Roman times, produced several engraved gems that contain a number of swordlike forms believed to be comets (Fig. 24). They are carved in a cosmic mix of the constellations, stars, sun, and moon, which are depicted above the sun-god Mithra as he sacrifices the sacred bull.

Several other ancient cultures represented comets. The Chinese, who were

Fig. 24. *Gem of the Astral God Mithra.* Udine, Civici Musei e Gallerie di Storia e Arte. Several comets are represented in the sky; the most easily identifiable are the sword-shaped variety.

advanced astronomically, recorded 272 comets between 611 B.C. and A.D. 1621, although the art of the Far East, curiously, contains almost no images of comets. It has been suggested that Central American tribes may have represented comets, and both the Aztecs and the Mayans were sophisticated in astronomical matters. Native American Indian depictions of comets have also been identified.

In the Judaic tradition, even though the Hebrews turned away from the study of astronomy (most likely because of its religious association for the neighboring Mesopotamians, whom the Hebrews considered idolatrous), many cometlike forms appear in literature. The Old Testament fairly abounds in suggestive images, such as the pillars of fire and smoke that guided the Israelites. An even more concrete image occurs in Chronicles (21:16): "And David lifted up his eyes and saw the angel of the Lord stand between the Earth and the Heaven, having a drawn sword in his hand stretched out over Jerusalem."

The New Testament contains even more graphic descriptions of both comets and meteors, especially in the visions of John the Evangelist in Revelation. With the advent of Christianity, pagan content often wore Christian clothing, so only the context of comet imagery changed. Popular belief that the fate of nations and individuals was ruled by the stars and planets remained, and comets, along with other celestial phenomena, became signs of God's wrath or approval. With Christianity, superstition rather than knowledge increased as the Middle Ages descended over Europe after the fall of Rome. But Arab astronomers kept the fires of astronomy burning in their translations of ancient texts. These Moslem treatises were, in turn, gradually translated into other languages, thus contributing to the general secular revival of classical learning during the Middle Ages and Early Renaissance.

The majority of comets from the Middle Ages are encountered in manuscript illuminations, for learning and erudition were centered in the Church and to a lesser extent in the courts. Many of these images are only references embroidering on ancient texts or on the writings of men such as Isidore, Bishop of Seville (born ca. A.D. 560). Due to the stylizations of the age, most of the comets depicted are abbreviated and schematic. Many are so simple, in fact, that they look like ordinary stars, which also may be the case with many stars depicted in Ancient and Early Christian times, wherein some images can only be identified as comets by accompanying inscriptions. One example, a large red star with eight rays found in the margin of a manuscript, is boldly identified as DE COMETIS STELLA; another is described as "a star which appears to fall and does not fall"; while a third is depicted as a circle inscribing a man's head from which issue eight rays of light.

Comets also appear in Christianized medieval astrological cycles. A case in point is the unusual personification of a comet carved on the archivolt of a thirteenth-century Italian Romanesque cathedral (Fig. 25), where it assumes a striking prominence second only to the sun and moon that flank the bless-

Fig. 25. Attributed to Niccolò, *Zodiacal Cycle* and detail of the central section of the archivolt, central west portal, Piacenza Cathedral, ca. 1222. The comet, inscribed *STELLA COMETA*, is personified by the stylized Romanesque profile inscribed in a circle at the right.

ing hand of God. The image is inscribed STELLA COMETA, a frequent designation at this time. It is balanced on the other side of the archivolt by a similar depiction identified as STELLA, which symbolizes all the stars of the heavens. The comet is differentiated from the star only by its inscription and the peculiar curling form of its hat. On each side of the arch, below the five central symbols of the heavens, are the signs of the zodiac and a trumpeting angel.

The thirteenth century was rich with the visual stimulations of great comet apparitions, among them Halley's Comet of 1222 and the Comet of 1264, which was one of the finest ever recorded, whose tail is said to have spanned over ninety degrees. The Italian monk Ristoro d'Arezzo related quite precisely that the Comet of 1264 rose at three o'clock in the morning "as large as a mountain." The Comet of 1299 was also spectacular and seems to have engendered more objective observations.

On the threshold of the fourteenth century, comets began to be observed more objectively for their physical properties. This new spirit was motivated in part by the desire to discover physical bases for astrological beliefs. It took the eyes of a painter, Giotto di Bondone, to shatter the established visual and literary conventions and to create the first convincing "portrait" of a comet, the 1301 apparition of Halley's Comet (Figs. 15 and 16).

During the humanistic fifteenth century, when the concept of individualism was reborn, the general populace still believed that all comets were the same object sent again and again to reveal God's will. If no comet apparition coincided with a pivotal event, people literally "saw stars" and invented one. They also hopelessly confused comets with meteors. It is significant that the earliest convincing representation of a meteor is in fact a meteor shower, which by its multiple nature is easily distinguishable from a comet. This manuscript painting (Fig. 26) was rendered for the Duc de Berry by the Limbourg Brothers, Flemish artists who settled in France and became pioneers in this age of burgeoning naturalism. Three gilt meteors appear in the background of this scene, depicting Christ's seizure in the Garden of Gethsemane, where they dramatically and symbolically pierce the enshrouding darkness. The meteors, which are noticeably absent in the illustration's textual source (John 18:4–6), create the profoundly religious but ominous mood and serve to foreshadow, like a comet, the tragic death of the King of Kings.

The fifteenth century—the age of Toscanelli and Leonardo da Vinci—was a century of contradictions. Despite a new critical spirit, many of the old superstitions persisted, as shown in a page of the *Beauchamp Pageants* (Fig. 27), recording events of 1403 during the rebellious wars in Wales (which began in 1402). The unusually bright daylight comet of 1402, blamed for the plague in the same year, is drawn at the upper right of this folio. Because it was indeed a daylight comet, the artist was accurate in including it during the battle. The delay time between 1402 and 1403 may explain the comet's curious placement in a separate section, a device that also serves to emphasize its significance. The same comet appears in a more stylized treatment on a page of another manuscript in the Biblioteca Casanatense, Rome, where it hovers over the fictional town of Aurimons, symbolic of the Church. Here it comprises part of a dark millenarian riddle; the text of the manuscript explains that the comet brings horrific catastrophes in its wake, similar in spirit to the panic recorded in the year A.D. 1000.

Astrological ideas retained a certain hold during the fifteenth century but became secularized and were often illustrated in a decorative manner for wealthy patrons. A case in point is an elegant manuscript thought to have been executed between 1456 and 1474, perhaps precipitated by the 1456 apparition of Halley's Comet (Fig. 28). Its fanciful comets ultimately descend from ancient types, especially the middle example shown here, which is in the shape of a sword.

The sixteenth century was undone by apparitions in the sky. The age of Copernicus, Luther, Michelangelo, Sir Thomas More, Shakespeare, and Tycho Brahe was a century of extremes, vacillating between the darkest superstitions and the promise of a new scientific age. While it brought Martin Luther's condemnation of astrology (he called comets "harlot stars"), the fear of retribution and guilt aroused by the Reformation only served to fuel the spread of rampant superstition on the popular level, and the widespread use of the printing press aided in the dissemination of such inflammatory material. On the other hand, this century, obsessed with collecting specimens and cataloging, witnessed the stirrings of the rebirth of science and astronomy. Copernicus's heliocentric theory, fully articulated in his *De Revolutionibus Orbium Coelestium* of 1543, was published earlier in a less complete form and invalidated Ptolemy's geocentric scheme, although conservative factions, such as the Church, were slow to accept this radical theory and its implications.

Representations of comets during this schizophrenic century tend to belong to a backward-looking tradition, best exemplified by the bold, contemporary scandal leaflets called broadsides. But other documents, such as

Fig. 26. The Limbourg Brothers, *Christ in the Garden of Gethsemane*, *Les Très Riches Heures du Duc de Berry*, Fol. 142r, 1416. Chantilly, Musée Condé.

Fig. 27. *The Comet of 1402*, *The Beauchamp Pageants*, Cotton Ms. Julius E. iv. Art 6, Fol. 3v. London, The British Museum.

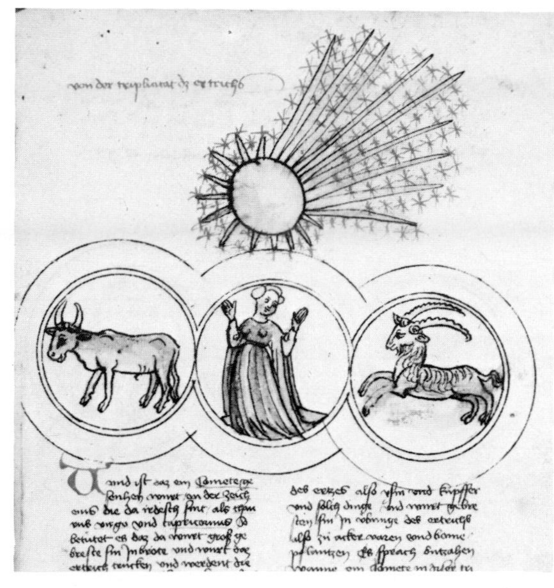

Fig. 28. *Three Folios of an Astrological Manuscript Showing a Comet in Virgo, Gemini, and Scorpio*, Ms. Pal. Lat. 1370, Fols. 124r, 129r, 126v, 1456–74. Rome, Biblioteca Apostolica Vaticana.

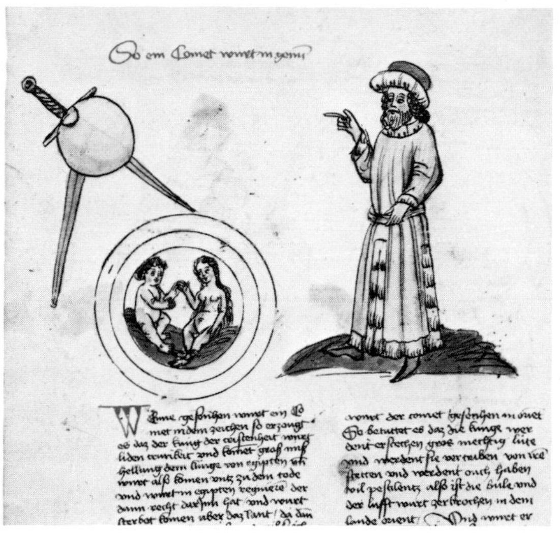

Apian's treatise of 1540, were scientific advances (Fig. 30). Illustrated town chronicles in manuscript and printed form, such as the *Nuremberg Chronicles* of 1493, embody the Renaissance's ambivalent attitudes toward comets (Fig. 13). While they aim to be objective in reporting natural phenomena and synthesizing earlier historical accounts, they frequently lapse into verbal and visual rhetoric, to say nothing of hyperbole. One more or less objective account is given on a page of the *Lucerne Chronicles*, ca. 1508–13, describing and illustrating the meteorite that fell to Earth at Ensisheim in 1492. Another page of the same manuscript, which illustrates two comets that supposedly appeared in 1472 (Fig. 29), lists the dire events caused by those comets. Only one comet is officially recorded for 1472, but since the chronicler notes that one appeared at midnight while the other was visible before sunrise, he may have observed the same comet twice, before and after perihelion. The manuscript also has an illustration of the 1456 apparition of Halley's Comet (Fig. 17), where it is also depicted twice, as well as an illustration of the Comet of 1506 and a supposed comet apparition in 1400.

Despite occasional lucid scientific forays and scholars' resuscitation of ancient teachings, the dire times of the Reformation, with its religious wars,

Fig. 29. Diebold Schilling, *The Comet(s?) of 1472*, *Lucerne Chronicles*, Fol. 77r, ca. 1508–13. Lucerne, Zentralbibliothek. The two apparitions look more like exploding celestial cannons than comets. Below, the plague has already claimed a victim, and destructive forces have been unleashed.

ASTRONOMICVM CAESAREVM

DECIMO QVARTO AVGVSTI DIE ſecunda Cometæ obſeruatio peracta eſt.

¶ Altitudo Cometæ ſupra horizontem gra. 8 mi. 29
¶ Azimuth Cometæ ab occaſu Septen. verſus gr. 45 mi 22
¶ Altitudo extremitatis caudæ graduũ 23 mi. 18
¶ Azimuth extremitatis caudinæ Septentrio. gra. 57 mi 38

Hæc ſequentia ex obſeruatione conſurgunt.

Latitudo Cometæ gra. 23 m̃ 2. Locus Cometæ gra. 23 mi. 39 ♌. Declinatio Come gr. 35 m̃. 12 Sept. Aſcẽſio recta Co. gr. 155 m̃ 5 Co. Mediat cœlũ ho. 11 añ. Amplitudo. or. & occ. 60 gr. 27 m̃ Sep. Occa. Come. hora 9 m̃ 32 poſt me. Come. occidit cũ 23 gra. ♏ Diſtantia Come. à Sole gr. 23 m̃ 40. Ortus Come. hora 2 mi. 28. Loc⁹ extremi. caudæ gr. 19 m̃ 18 ♌. Latitu. extre. gr. 39 m̃ 45 Sept.

Situs Cometæ occaſus tempore. Situs Cometæ ortus tempore.

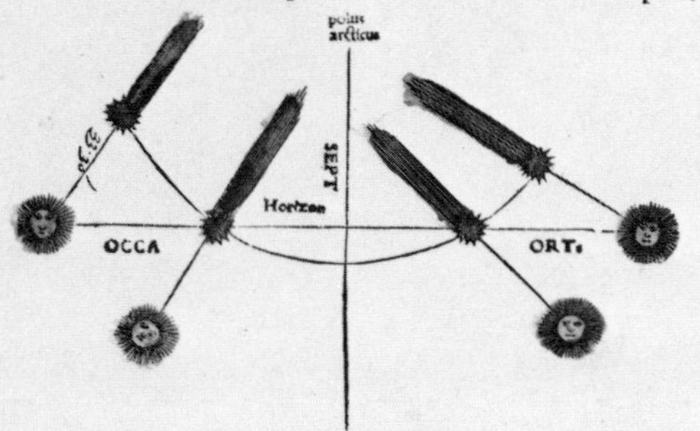

DE OCCVLTATIONE ET APPARITIONE Cometæ. Caput decimumſextum.

TAMETSI Cometes 13 auguſti die mihi cõſpectus ſit primo, non deſunt tamen qui eundem ſe vidiſſe ſexto ſeptimoq; affirment, verè autem omnibus ab occidẽte paucis ab oriente viſus eſt, & profecto fieri non poteſt, quin ita factum ſit, licet non omnibus apparuerit, non enim ita multis poſt ortum ſuum diebus videri mane deſit, ea de cauſa, quod ſubinde magis ac magis coſmicum in ortum feſtinarit. Nam die 18, vt videre in ſequentibus eſt, oriri cum ipſo Sole cępit, quocirca propinquor erat iam Soli 13 die quam vt cerni potuerit, ortum parante iam Sole. Inde factum eſt,

DECIMO QVINTO AVGVSTI OBſeruatio tertia facta eſt.

Fig. 30. *Halley's Comet of 1531*, detail of a page from Petrus Apianus (Apian), *Astronomicum Caesareum*, 1540. This page illustrates Apian's observation that a comet's tail always points away from the sun, an idea also postulated by Fracastoro.

peasant-class wars, and absolutism, produced a generally apocalyptic climate. They thus serve as a barometer of the times.

> As with its bloody locks let loose in air,
> Horribly bright the comet shows whose shine
> Plagues the parched world, whose looks the Nations scare
> Before whose face States change and Powers decline,
> To purple tyrants all, an auspicious sign.
>
> (Torquato Tasso, *Jerusalem Liberated*, VII)

Especially in northern Europe, people turned with impassioned zeal to reading as omens such natural events as unusual births, hailstorms, the northern lights—and most notably, comets. Woeful tales of God's merciless wrath were collected and circulated in volumes that chronicled all calamities and monstrosities since the beginning of history. Lycosthenes' large volume *Prodigiorum ac ostentorum chronicon* (*A Chronicle of Prodigious Events*) may be the most amibitious of this genre. It features accounts of many comets, some fanciful, others real. In occasional bold woodcut illustrations (Fig. 38) that are repeated throughout the text, comets take many marvelous guises, such as swords, or even more exotic forms—a bloody shower of daggers, axes, knives, and hideous faces, as described by the French physician Ambroise Paré in

Fig. 31. Albrecht Dürer, *Melencolia I*, 1514, engraving. Even before Dürer's time, there existed an esoteric tradition wherein comets, as harbingers of negative forces, were associated with the god Saturn, who ruled over the state of melancholy.

Fig. 32. Matthias Gerung, *Apocalypse Illustration*, 1547, woodcut. It illustrates the comet-like form described in Revelation 9:1–3: "And the fifth angel sounded, and I saw a star fall from heaven onto the earth: and to him was given the key of the bottomless pit. And he opened the bottomless pit and there arose a smoke out of the pit . . . locusts . . . scorpions. . . ."

Fig. 33. Matthias Gerung, *Apocalypse Illustration*, 1547, woodcut. This print depicts the Star Wormwood from Revelation 8:10–11: "And the third angel sounded and there fell a great star from heaven burning as it were a lamp. . . . And the name of the star is called Wormwood."

Fig. 34. *The Comet Type Known as Argentum*, Ms. FMH 1290, p. 21, ca. 1587. London, The Warburg Institute. This positive comet has silvery rays resembling the complexion of Jupiter and prophesies a time of abundance and fertility. Fig. 35. *The Comet Type Known as Aurora*, Ms. FMH 1290, p. 17, ca. 1587. London, The Warburg Institute. This comet's complexion is red, and in Germany it was said to bring famine, wars, and the burning of houses.

1528; or an Arabian comet (Fig. 38) that looks like a spaceship straight from the art-deco sets of *Flash Gordon*. It is difficult to comprehend that this volume was published seventeen years after Apian's progressive treatise.

As part of the revival of humanistic learning, classical astronomy was once again of interest in western Europe. The constellations were organized into beguiling star maps to chart the course of comets across the sky. Albrecht Dürer published forty-eight wood engravings of classical constellations in 1515, and depicted several comets. Not only did he position the saturnine figure in his *Melencolia I* (Fig. 31) under the alchemical aegis of a comet—a harbinger of change and disasters—but he also employed meteors and cometlike forms to visualize certain passages of Revelation in his woodcut Apocalypse series (Fig. 5). Both Dürer and Matthias Gerung, who followed suit in 1547 with his own more specific Apocalypse series (Figs. 32 and 33), relied on, yet enriched, a well-established tradition.

The nine or ten ancient types of comets supposedly described by Ptolemy and subsequent commentators were also revived and given contemporary illustration in several manuscripts of this period. The most sumptuous ex-

Fig. 36. *The Comet Type Known as Pertica*, Ms. FMH 1290, p. 39, ca. 1587. London, The Warburg Institute. This type supposedly characterized Halley's Comet in 1531. Reputedly, it looks like a column when seen in the west, like a glowing star in the east. Fig. 37. *The Comet Type Known as Veru*, Ms. FMH 1290, p. 35, ca. 1587. London, The Warburg Institute. A comet of this type was supposedly seen in conjunction with the A.D. 66 destruction of Jerusalem. This would be Halley's.

ample shows the various comet types hovering above landscapes with contemporary features in full-page illustrations (Figs. 34–37). The accompanying textual accounts describe their devastating terrestrial effects, summed up quite well in *La Semaine*, Guillaume de Salluste du Bartas's sixteenth-century epic poem about the Creation (see half-title page).

The sixteenth century kept the new printing presses busy with treatises on comets. One example is Nicolaus Pruckner's work of 1532, probably inspired by the 1531 apparition of Halley's Comet, as well as by the Comet of 1532 (see illustration on page vii). Another illustrated comet treatise by Mathias Brotbeyel of the same year features a skeleton with a bellows blowing the comet along its destructive course.A large watercolor from a commonplace book, a collection of transcribed passages, preserves a naively rendered but powerful image of the Comet of 1532 (Fig. 42), which was carefully observed by Apian and the Italian poet-astronomer Girolamo Fracastoro, who claimed it was three times as bright as Jupiter. This vivid image is also significant in that it shows an attempt by an untrained hand to record an empirical observation of a natural phenomenon.

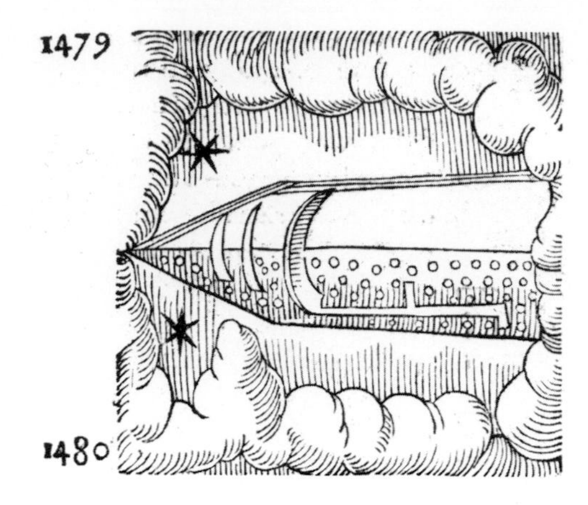

Fig. 38. *Comet Over a City and The Comet Seen in Arabia in 1479*, from Conrad Lycosthenes, *Prodigiorum ac ostentorum chronicon*, 1557.

The next apparition to attract a great deal of attention was the Comet of 1556, effectively preserved in a broadside (Fig. 39). These sheets flooded the countries of Europe, especially Germany, and luridly proclaimed each event in the manner of today's tabloids. These broadsides had a significant influence on people's lives, for they were the closest equivalent of modern empirical reporting on current human events. Excitement over the Comet of 1556 resulted in the publication of additional treatises, some of which also debated the relevance of comets to astrology, as well as the validity of astrology itself. By the end of the century, astrology's authority was vanquished for the intelligentsia, though it would continue to hold sway over the general population.

Fed by the epidemic proportion of comet fever, artists and writers continued to portray comets in traditional contexts during this turbulent period. Only fourteen verifiable apparitions were witnessed, but scores were reported. Curiously, the Star of Bethlehem was only rarely represented as a comet in Italy, where after Giotto a watered-down tradition persisted. In German art, one finds instead huge bursts of blinding light that might be interpreted as comets without tails but are more properly termed *Wundererscheinungen*, "wondrous shining appearances." Heraldry, certainly the last bastion of medieval tradition, continued to employ comet imagery. Italians seem to have had a penchant for using comets in personal emblems (*imprese*) as well as family coats-of-arms, probably to indicate a "star-crossed" personality trait in an individual. During the sixteenth century, comets in a heraldic context were expressed on a grander scale than ever before. For example, they occur on the sculptured Doric frieze that decorates the Palazzino Gambara of the Villa Lante at Bagnaia, as well as in the frescoed loggia below, as a personal *impresa* of the cardinal. They also appear on the Doric Frieze of the present Biblioteca Hertziana in Rome, signifying the star of Pope Urban VII

Fig. 39. Herman Gall, *Of a Comet and Two Earthquakes in Rossana and Constantinople, 5 March 1556*, colored woodcut.

Fig. 40. Jiri Daschitzsky, *Broadside of the Comet of 1577 Seen at Prague*, published by Peter Codicillus, woodcut.

Fig. 41. *The Comet of 1577*, from *Secaatname*, an illuminated Turkish manuscript. Istanbul, The Central Library of Istanbul University.

Fig. 42. *The Comet of 1532*, page from a commonplace book, watercolor, 8¾ x 11½ in. London, The Science Museum Library.

under whose pontificat the artist, Federico Zuccaro, bought the lot to build this house.

Cometology as a science was born during the sixteenth century. As previously mentioned, Apian correctly postulated that a comet's tail always points away from the sun in his treatise (Fig. 30). But by far the greatest pre-telescopic observational astronomer was the Dane Tycho Brahe (who wore a golden nose as the result of losing his own in a duel). He had an unusually well-equipped personal observatory which he named Uraniborg. Among his many accomplishments, Brahe issued an accurate star catalog, discovered the Comet of 1582, and finally proved that comets were distinct objects in space rather than the threatening atmospheric phenomena that Aristotle had described. Much of Brahe's work concerned the bright daylight Comet of 1577 (Fig. 40), which he correctly positioned as orbiting the sun (which, however, he mistakenly described as orbiting the earth).

The Comet of 1577 was also discussed in such a flood of treatises that it took one scholar over a hundred pages to compile a bibliographical list of them. Some of these treatises were enlightened, such as one by the Spanish physician Marcellus Squarcialupus, who ridiculed those who were afraid of comets. He was especially annoyed with members of the clergy who used the appearance of a comet to frighten the ignorant, and stated that most of the ills blamed on the poor stars were actually caused by men. The notorious Comet of 1577 caused so much excitement that it also inspired such scientific achievements as the founding of the Istanbul Observatory, which may have been instrumental in the creation of a group of unusually vivid Turkish comet illustrations, among them the one shown in Figure 41.

The importance that sixteenth-century people attached to such astronomical events is also reflected, though to a much lesser extent, in monumental painting. While eclipses and lunar phases were more commonly depicted, cometlike forms are rather rare. Raphael painted one in the background of his *Madonna of Foligno*, executed in 1512 (Fig. 44), wherein a meteor or fireball is shown falling on a house in the Venetian-inspired landscape. (It is interesting to note that, as in Dürer *Melencolia I* (Fig. 31), the comet is accompanied by a rainbow.) A heavenly missile actually did hit his patron Sigismondo de Conti's house, thus both singling him out as an exceptional individual and foreshadowing his death in 1512. As Shakespeare wrote in *Julius Caesar* (II:ii), "When beggars die, there are no comets seen,/The heavens themselves blaze forth the death of princes." Shakespeare employed many other comet allusions, including one describing its "crystal tresses" (*Henry VI*, I:i).

During the sixteenth century, well known as an innovative era, painting graduated from the status of a craft to become one of the fine arts. Indeed, it was in this new age that Michelangelo was christened "the divine one." With this rise in artists' status came a greater freedom to invent images, so it is not surprising that the art of this century abounds in novel and imaginative contexts for comets. An example is Lucas van Leyden's *Lot and His Daughters* (Fig. 45). The artist set this Old Testament scene against the backdrop of Sodom's destruction, presided over by an eerie comet at the upper left and a rain of meteors at the upper right. Literary sources contemporary with this painting also mention a comet sent by God to signal the destruction of Sodom and Gomorrah. In a similar spirit, a woodcut illustration of the fall of Troy for Virgil's *Aeneid* (1502, edited by Sebastian Brant) includes the menacing form of a hairy star, much lamented by the poet in other passages, hanging over the burning city. The comet is ironically welcomed as the sign sent by Jupiter to indicate that Anchises and Aeneas should leave Troy. Whereas Anchises rejoices in its appearance, the comet ironically also signals the death of Priam and the destruction of Troy. Virgil, who frequently describes celestial imagery (including comets, usually employed as sinister signs) throughout his *Aeneid*, *Georgics*, and *Eclogues*, was certainly conscious of this double meaning.

A certain aesthetic distance from the burgeoning young sciences practiced under the auspices of the Medici Court in Florence, and perhaps a note of humor, was demonstrated by Giorgio Vasari and his followers in a drawing

Fig. 43. *Montezuma Transfixed by a Comet in 1519–20*, from Diego Duran, *Historia de las Indias de Nueva España*, written 1574–81, published 1867–80.

(in the Uffizi, Florence) for a costume designed for the personification of *La Cometa* to appear on a float in a Medici marriage fete. The image echoes lines from Shakespeare's *Henry VI* quoted above, as well as Edmund Spenser's *Faerie Queene*:

> All as a blazing starre doth farre outcast
> His hearie beames, and flaming lockes dispread,
> At sight whereof the people stand aghast;
> But the sage wisard telles, as he has red,
> That it importunes death and doleful drerihed.

When explorers opened the New World to Europeans in the sixteenth century, they found among the natives a fear of comets that resembled their own antiquated superstitions. The Incas of Peru regarded comets as an intimation of divine wrath from their sun-god and believed that Halley's apparition of 1531 had heralded Pizarro's invasion. Several Spanish histories relate that Montezuma, ruler of the Aztecs, was so frightened by the appearance of a comet that, on the advice of court astrologers, he refused to act while it hung in the sky (Fig. 43). Cortés thus easily subdued the Aztec empire and was himself regarded as the blond-haired god of Aztec myth.

While superstitions continued to haunt popular ideas about comets during the sixteenth century, the serious scientific inquiries into their nature pushed pretelescopic observation to its limits. The next step was taken after the turn of the century by Galileo, who in 1609 learned about the magnifying lenses that he would later employ in his first telescopic observations.

Fig. 44. Raphael, detail of *The Madonna of Foligno,* 1512, oil, 126 x 77¼ in. Rome, Pinacoteca Vaticana. The fireball which actually hit the patron's house in Foligno marked him as an unusual individual and presaged his death.

Fig. 45. Lucas van Leyden, *Lot and His Daughters,* ca. 1509, oil, 22⅞ x 13⅜ in. Paris, Musée du Louvre. The painting illustrates the fall of Sodom with "brimstone and fire out of heaven" (Genesis 10:24), sometimes interpreted by sixteenth-century writers as a comet.

Three

The Comet Crisis and the Birth of Telescopic Astronomy

The quest for truth and authenticity, which is associated with the Reformation in the sixteenth century, culminated in the following century with the birth of modern science and its stress on verification by experience. After nearly fifty years of Counter-Reformation austerity and purification, the Catholic Church decided to combat the Protestant challenge in a positive manner, by stressing humanity and sensual immediacy. During this period, mysticism and pessimism were replaced by empiricism and optimism in both art and science, as exemplified in the works of Bernini, Rembrandt, and Galileo. Reasoned mathematics became almost a religion to some of the finest thinkers of the century—such as Descartes, Newton, and Halley—and helped to greatly advance astronomical endeavors. Only Kepler and Newton were distracted by less empirical fields (the former by astrology and the latter by alchemy), but not before they had made monumental contributions to modern science. All these developments brought the study of comets to a critical point.

The seventeenth century was an age of heretofore unparalleled progress in scientific investigation. The wealthy mercantile Protestants of Holland and the post-Reformation Catholics of Italy led the way. Light can be seen as a leitmotiv of this new era; it suffused all European art and architecture, while the interest in optics and the study of light led to an improvement of lenses and a more intense study of the heavens. Numerous individuals in England, Holland, and Italy experimented with lenses and telescopes, but it is Galileo to whom credit is given for the invention of the telescope and its profound application to astronomy. Galileo, who observed the two comets of 1618 as well as a supernova in 1604, was a brilliant example of the new scientific thinker. Not only did he adopt the Copernican heliocentric system, but he also began to research astronomical problems as purely mechanical ones, without the metaphysical obstructions of earlier attitudes.

Fig. 46. *Polish Print of Astronomers Watching a Comet*, 17th century, woodcut. They are using an astrolabe, a quadrant, and a compass poised atop a globe with the zodiacal band.

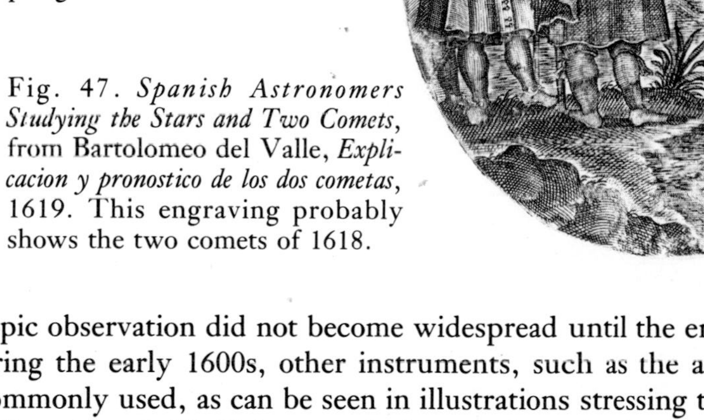

Fig. 47. *Spanish Astronomers Studying the Stars and Two Comets*, from Bartolomeo del Valle, *Explicacion y pronostico de los dos cometas*, 1619. This engraving probably shows the two comets of 1618.

But telescopic observation did not become widespread until the end of the century. During the early 1600s, other instruments, such as the astrolabe, were more commonly used, as can be seen in illustrations stressing the act of observation (Figs. 46 and 47). Telescopes had to await perfection by such innovators as the Dutchman Christian Huygens, who improved the instrument and thus helped advance astronomy. Huygens also recognized the necessity for accurate timekeeping in conjunction with scientific observation, and hence incorporated Galileo's pendulum into his clocks. Finally, in 1668, Newton constructed a reflecting telescope, the ancestor of the large modern telescope.

At the same time, several of the absolute monarchs of the age founded national observatories to house the new, more powerful optical equipment. Two of the most famous, those of Paris and Greenwich—commissioned respectively by the Sun King, Louis XIV, in 1667 and by Charles II in 1676—enabled some of the greatest scientists of the day to observe the course of comets as was never before possible.

It was a critical time for theories about comets. Enough data was known to create a state of crisis. Halley and Newton, the proper fathers of cometological studies, forged their pivotal theories during this epoch (see Chapter 1). It was, therefore, especially appropriate that the apparition of Halley's Comet in 1607 was the first comet of the century.

In 1618, two comets illuminated the skies (Fig. 47). The second one was the first brilliant comet of the century, and its tail reached a maximum of seventy degrees, dramatically splitting into numerous starlike fragments. Kepler observed this comet, as did Milton, who was just a boy. Milton con-

Fig. 48. *Comet Types*, from Alain Mallet, *Description de l'univers*, 1683.

Fig. 49. *Austrian Print of the Comet of 1665*, etching.

tinued to be haunted by it and frequently alluded to comets in his writings, perhaps because after losing his sight he realized that it was the only comet he would ever see.

> So spake the grisly Terror, and in the shape
> So speaking, and so threatening grew tenfold
> More dreadful and deform. On the other side,
> Incensed, with indignation, Satan Stood
> Unterrified and like a comet burned. . . . (*Paradise Lost*, II)

New scientific knowledge was slow to change the traditional representation of comets. For example, a comet occurs in a print, after a painting by one of the Caracci, in which the Star of Bethlehem once again takes the form of a comet. Broadsides proclaiming prodigious events were also issued, but less frequently and with an attempt to distinguish more closely between the types of comets observed. The subtle change of emphasis is echoed in people's willingness to accept comets as natural phenomena with terrestrial effects rather than as divine messengers. Later in the century, ancient types of comet shapes were again illustrated, but in a more up-to-date format arranged in an attempt to classify them systematically (Fig. 48).

Between 1618 and 1665, four comets appeared, but none struck terror in the hearts of Europeans until the Comet of 1665 lit up the sky (Figs. 49 and 50) and the plague descended on the Continent. It motivated the Englishman John Gadbury to publish *A Discourse on the Nature and Effects of Comets or Blazing Stars*, whose frontispiece also cites the Comet of 1664 and quotes du Bartas's rhyme (see half-title page).

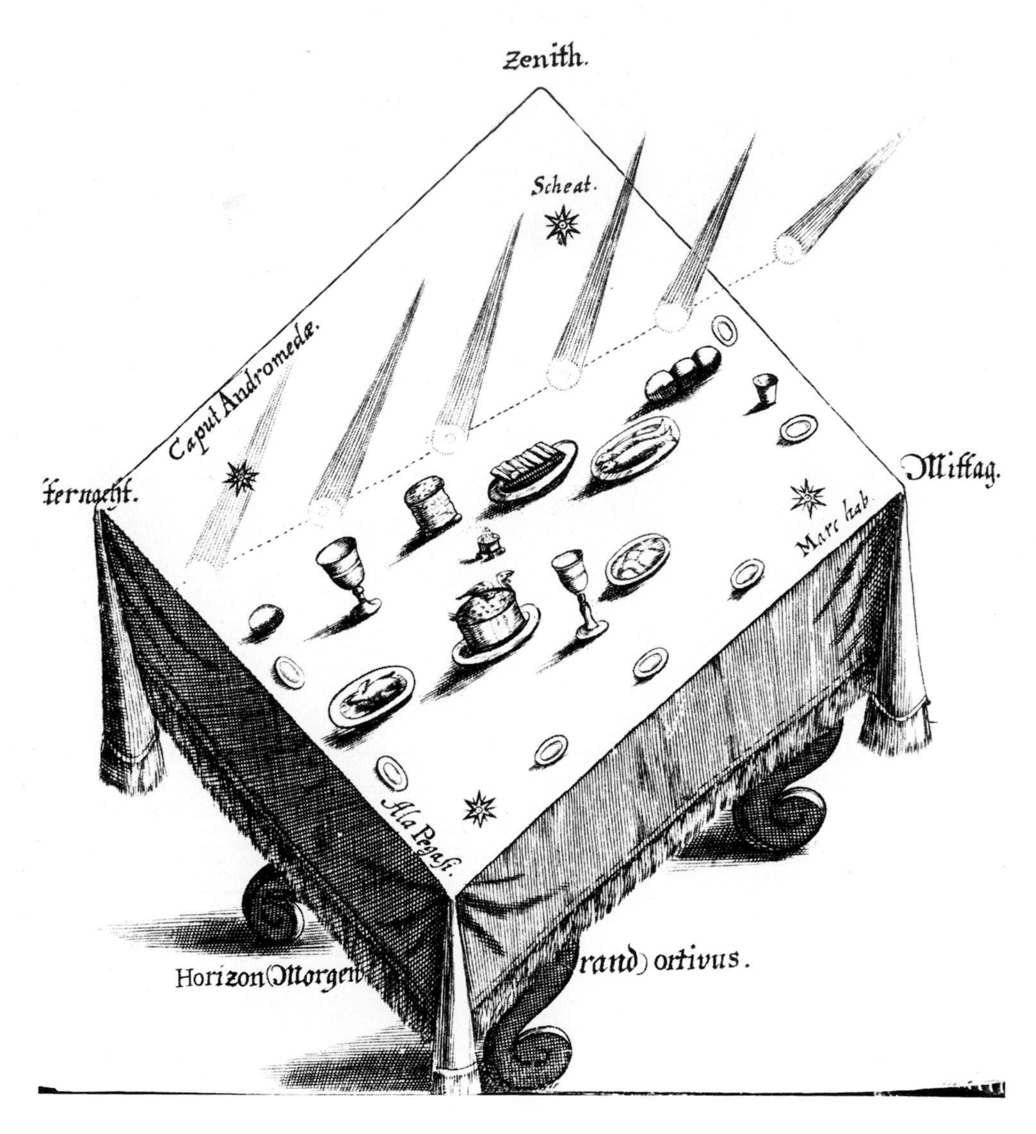

Fig. 50. *The Comet of 1665*, from Christoph Richter, *Berichtendes . . . Cometen.*

The image of the Comet of 1665 may have been preserved by the Dutch artist Herman Saftleven in his strange canvas depicting an *Allegory of the Events Following the Execution of Charles I* (Fig. 52). The overall tone of the painting is set by the menacing comet, which is correctly represented in its gossamer luminescence and potent in its symbolism. At the right, a painting within the painting portrays the execution on January 30, 1649, of Charles I

Fig. 51. Andreus Stech, frontispiece from Johannes Hevelius, *Cometographia*, 1668.

of England in front of the inaccurately rendered Banqueting Hall of Whitehall Palace in London. The masked executioner-fox regarding the painting may be Oliver Cromwell, whom the Dutch represented as the Antichrist as early as 1656, and as the embodiment of violence as late as 1680. The slavering hound to his right may have been intended to caricature Thomas Fairfax, the commander-in-chief of the parliamentary army during the Civil War. This hound holds a ribbon in its teeth, inscribed *Per Tirannia*, which is attached to the discarded scales and sword of justice. Under the dog's feet are scattered works by Luther, Calvin, and others. Various weird, often composite, figures surround or perch at the edges of the painting, such as the fox clutching a dagger with a Latin inscription that is only partially legible. To the left are rendered the evils unleashed by the execution. In the center, Mars, the belligerent god of war, is shown in a bestial act. Further to the left is a building engulfed in flames, before which is perhaps the Mouth of Hell, spewing forth horrific evils such as Famine, Ignorance, and Vice. Nearby, a pig vomits inscribed documents; while many of the inscriptions are effaced, one paper is clearly labeled *false* in Latin. In the right background, the city of London burns uncontrollably. The view here closely corresponds to other seventeenth-century views of the city, but according to accounts of the 1666 fire of London, the wind blew from the east, while here it blows from the west, which may mean either that the painting was prophetically executed before the fire of 1666 or that the artist did not know about the nature of the wind.

At times, the naturalism of seventeenth-century artists and writers came to loggerheads with the old, deeply ingrained beliefs about comets. England's poet laureate, John Dryden, humanly confessed that he could not decide which of the many theories about comets to believe. Yet despite the uncertainty about their nature, comets were a very popular image during the century's flowering of poetry, as for example in the verse of John Donne, who composed a song beginning "Goe and catch a falling star. . . ." And, in his poem "To the Countess of Huntington," he wrote:

> Man to Gods image; *Eve*, to mans was made,
> Nor finde wee that God breath's a foule in her. . . .
>
> Who vagrant transitory Comet see,
> Wonders, because they are rare; But a new starre
> Whose motion with the firmament agrees,
> Is miracle; for there no new things are;
>
> In woman so perchance milde innocence
> A seldome comet is, but active good
> A miracle, which reason scapes and sense,
> For, Art and Nature this in them withstand
>
> As such a starre, The Magi led to view. . . .

Fig. 52. Herman Saftleven, *Allegory of the Events Following the Execution of Charles I*, ca. 1649–66, oil, 22 x 32 in. Petworth House, The National Trust.

Some of the most lavish, large-scale multivolume comet treatises ever published were produced during this era, and two of the most important were precipitated by the Comet of 1665. The first, entitled *Cometographia*, was written in 1668 by the German astronomer Johannes Hevel, known as Hevelius, after his shorter *Prodromus cometicus* of 1665. It was the first systematic account of past comets and was done in collaboration with Hevel's wife, Elizabeth. Its frontispiece (Fig. 51) reveals the century's puzzlement about comet orbits before Newton and Halley's ideas were formulated. All three astronomers depicted in the print hold diagrams of their incorrect theories. In the center, the author-astronomer points to the illustration of his theory, which modified Kepler's idea that comets travel in straight lines, as seen in the right-hand astronomer's drawing. Instead, Hevelius believed that comets travel in a slightly curved trajectory around the sun. (The frontispiece also demonstrates the century's preoccupation with scientific instruments, among them a telescope.) The second treatise, *Theatrum Cometicum*, first published in 1666 in Holland, was written by the Pole Stanislaw Lubieniecki. His opus is a universal history of comets (Fig. 53), from the biblical flood through the Comet of 1665, illustrated in a sumptuous double-page spread that charts the course of that comet against the constellations.

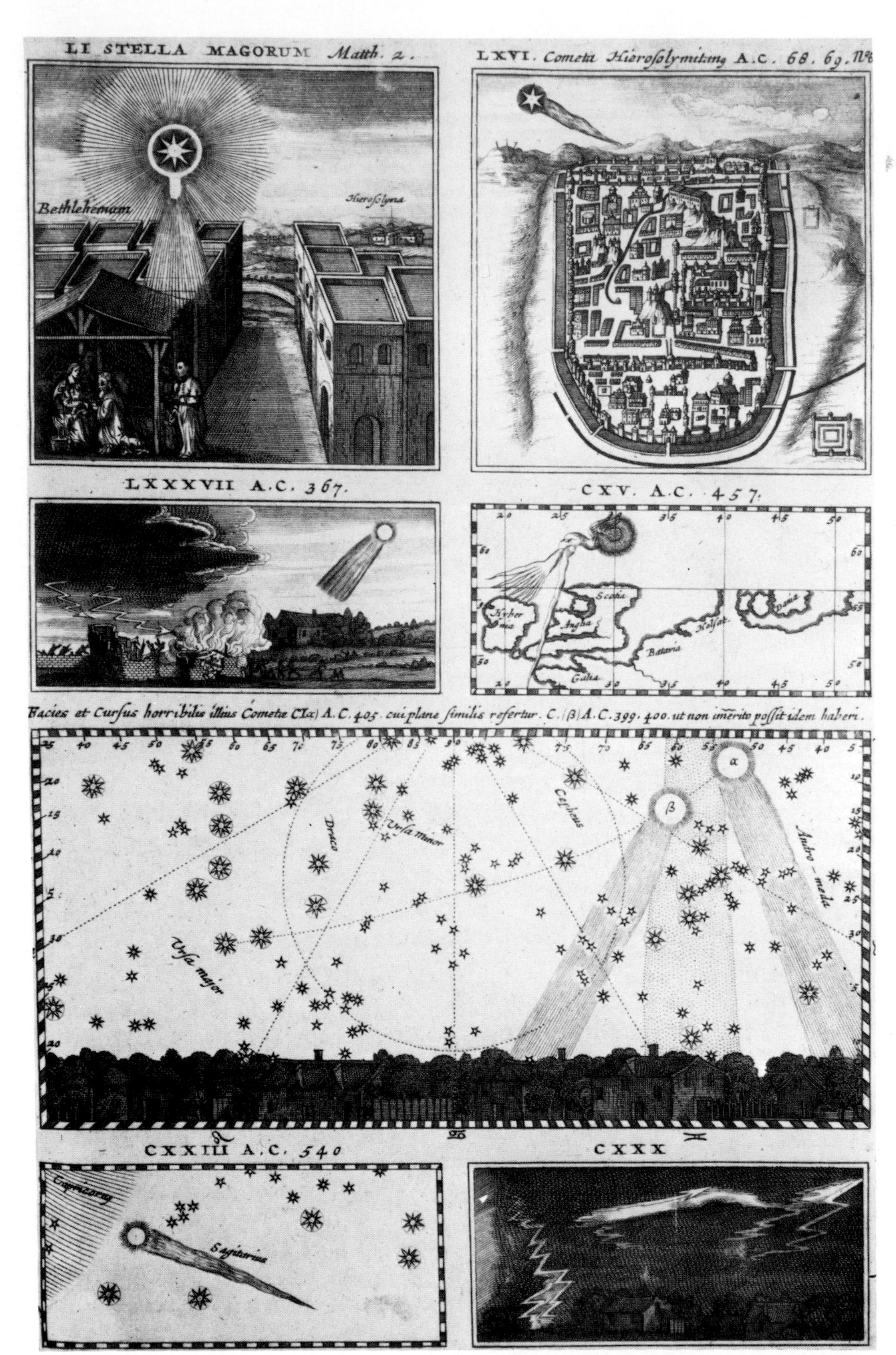

Fig. 53. Page showing the Star of Bethlehem and types of comets, from Stanislaw Lubieniecki, *Theatrum Cometicum*, 1681 edition.

Fig. 54. Johann Jacob von Sandrart, *The Great Comet of 1680*, 1681, etching.

The first telescopic discovery of a daylight comet in 1680 caused a flurry of astronomic activity (Fig. 54) and also revived old superstitions. While astronomers coolly gazed at the apparition through telescopes, an irrational element in society grew hysterical about the gigantic red-tinted comet, one of the most brilliant ever recorded. A naked-eye comet with a seventy-degree tail, it was first spotted before sunset on 14 November 1680 and last observed on 19 March 1681 by Newton, who concluded that comets are controlled by the sun's gravitation.

Newton's gravitational theory was a direct attack on all superstitions connected with comets, which during the seventeenth century grew to encompass witchcraft. Pierre Bayle, in his classic *Pensées sur la comète*, published in 1682, presented a scathing satirical attack on the bigotry and intolerance associated with comet superstitions. Bayle, a teacher in Rotterdam, was attacked in turn by the reactionary preacher Balthassar Bekker, who did believe in witchcraft. A similarly Calvinist mood prevailed in New England, where Increase Mather stated in a sermon of 1680, "Heaven's Alarum to the World," that comets are "God's sharp razors on mankind." The culmination of these and other such sentiments was the infamous Salem witch-hunt of 1692, in which nineteen persons were executed.

A cryptic German broadside also depicts the Comet of 1680 and supports

Fig. 55. *Broadside of the Comet of 1680 Fulfilling a Prophecy of Nostradamus,* engraving.

the wild rumor that an egg had been laid in Rome on which stars and a comet were represented (Fig. 55). This incident supposedly relates to an astrological prophecy attributed to Nostradamus. As in earlier years, comet medals were struck in Germany for the Comet of 1680; one displays a curved orbit, and another is inscribed: *This star threatens evil things. Trust in God who will turn them into good.*

In the debate that raged between the superstitious ideas of the past and the more progressive rational approach, the Comet of 1680 was a portent of enlightenment. Thomas Brattle observed the comet methodically at Harvard College in Cambridge, Massachusetts, and his results reached Newton in time to be incorporated into the *Principia*. Madame de Sévigné wrote that the giant comet had the most beautiful tail she had ever seen. Like Cardinal Mazarin's, her views reflect objective common sense and a disbelief that comets should frighten anyone. Bernard le Bovier de Fontanelle, a frequent guest at Madame de Sévigné's salon, spoofed astrology in his comedy *La Comète*. For the first time, comets were regarded with enough dispassion that they could be relegated to decorative motifs, as on the unusual clock by Joannes Hessler in the Poldi-Pezzoli Museum, Milan, probably inspired by the Comet of 1680, where each hour is separated by a comet. This was a particularly appropriate motif as clocks were an essential technical device to aid in accurate astronomical record-keeping.

Fig. 56. *The Comet of 1680 Over Beverwijk on December 22*, watercolor, 13½ x 9⅜ in. Rotterdam, Atlas van Stolk. The inscription terms it a "ghastly star."

Fig. 57. Lieven Verschuier, *The Comet of 1680 Over Rotterdam*, oil, 8⅞ x 12¾ in. Rotterdam, Historisch Museum.

Dutch artists, who were steeped in Protestant realism, seem to have been particularly successful in expressing the new scientific outlook. A case in point is Lieven Verschuier, whose painting *The Comet of 1680 Over Rotterdam* (Fig. 57) correctly depicts the important landmarks of the city. Two inscribed preparatory drawings attest that the canvas depicts the comet on the night of December 26 and that its magnificent tail has been cropped. There are a number of Dutch depictions of this comet, including a freer, more abstract watercolor that displays its huge tail that dominates the entire sky (Fig. 56). This bold, anonymous drawing is stylistically so progressive that it has intimations of nineteenth-century techniques. After 1680, eight more comets crossed the heavens before the close of the century, but with the exception of Halley's Comet in 1682 (see Chapter 1), none created the furor of the Comets of 1665 and 1680.

Most of the comet-oriented thought of the seventeenth century focused intently on critical scientific theories, such as those of Newton and Halley. Artists were less concerned with comet imagery than with more generalized depictions of light. Comets do, however, appear in novel and more varied artistic contexts than ever before, as in C. G. V. Amling's print, after Candid, *Autumn* (Fig. 58), where the comet is symbolic of a bountiful harvest and an auspicious vintage. This earthy context, typical of the sensual subject matter common in seventeenth-century art, expresses the prevailing theory that comets affected the weather, a belief that supposedly accounted for the superiority of wines made in comet years. Comets were fallaciously believed to cause higher temperatures and thus a higher sugar concentration in the grapes from which wine was made; this belief persisted until the nineteenth century.

Individuals of the time were also passionate about emblems and allegory; in Jeremias Falck's engraving *Astrology*, for example, comets or shooting stars function as attributes of the personification of Astrology. In another, more complex engraving of 1698 by Amling, also after Candid, *Night*, a comet once again hovers symbolically in the heavens as a naturalistic attribute of night and of the symbolic heritage of Western civilization.

Fig. 58. C. G. V. Amling, after P. de Witte (Candid), *Autumn*, 1698, engraving.

Fig. 59. David Roentgen, *Intarsia of Astronomy and Geometry*, 1779, inlaid wood, 26¾ x 20½-in. Munich, Bayerisches Nationalmuseum.

Four

Cometology Comes of Age

The telescope beckoned astronomers beyond the confines of our solar system, promising a universe more vast than previously imagined. During the eighteenth century, astronomical observations were made even more systematically and with increasing accuracy, due to such new sophisticated instruments as the precision chronometer, a device for keeping time. Sir Frederick William Herschel, Charles Messier, and Caroline Herschel (Sir Frederick's sister and the first female comet-hunter) were among the astronomers intensely dedicated to observation. Extraordinary progress in the fields of optics and physics also aided in the sighting of at least sixty-two comets during the century; during its last quarter, at least one comet was spotted each year. The interest in celestial observation became so widespread that by 1800 observatories began to multiply not only in Europe but throughout the world.

During the eighteenth century, cometology came of age. In 1705 Halley published his significant findings in the treatise *Astronomiae Cometicae Synopsis*, and then, sixteen years after his death, Halley's Comet returned on Christmas Day of 1758, precisely when Halley predicted it would. The return proved not only that comets could be periodic and that they travel in elliptical orbits around the sun, but also that the Newtonian system of gravitation was valid. Before that fateful event, Newton's ideas were accepted by very few people outside of Britain. This insular attitude was partly due to the difficult mathematical methods that Newton employed; they were geometric rather than the analytic methodology practiced on the Continent by, among others, the mathematician and philosopher Gottfried Leibniz.

The eighteenth century was also the age of the rococo style and the Enlightenment, which in the later part of the century led to the stoic morality of Neoclassicism. It was the era of Voltaire, Immanuel Kant, Wolfgang Amadeus Mozart and Denis Diderot's twenty-four-volume *Dictionnaire Raisonné*. It was also the time of Charles Messier, called "the comet ferret" by Louis XV in recognition of the fifteen or more comets he discovered. Comets continued to function as literary images, but during the eighteenth century they were often treated more dispassionately, as in Jonathan Swift's savage *Gulliver's Travels*, in which Swift satirizes, among other things, the superstitious fear of comets.

In the world of fashion, the extravagant imagination of the century was expressed in the wigs worn by members of high society. One was called "The Comet," perhaps in celebration of Halley and his posthumous triumph in 1758. Science itself became fashionable and even romantic in the last quarter of the century as seen in an inlaid wooden panel of 1779 by David Roentgen (Fig. 59)—which also demonstrates that the century was an age of fine craftsmanship. It is appropriate that the artist has depicted an astronomer observing a comet to represent the discipline of astronomy, since the work dates from about twenty years after Halley's vindication.

Archaic superstitions were still strong at the beginning of the century, when broadsides and book illustrations continued to extol the prodigious properties of comets. As the idea that comets were celestial objects rather than portents gained wider acceptance, another very basic fear—that a comet would collide with Earth—began to supplant old superstitions. In 1758, for example, John Wesley, the English theologian and founder of Methodism, used the return of Halley's Comet in a sermon to warn people to make their peace with God, lest a comet hit the earth and bring the Apocalypse. In 1773,

BOSTON: Printed and Sold by B. GREEN and Comp. in *Newbury-Street*, and D. GOOKIN, at the Corner of *Water-ſtreet, Cornhil.* 1744.

Fig. 60. *The Comet of 1744*, frontispiece from Mather Byles, *The Comet: A Poem*, 1744.

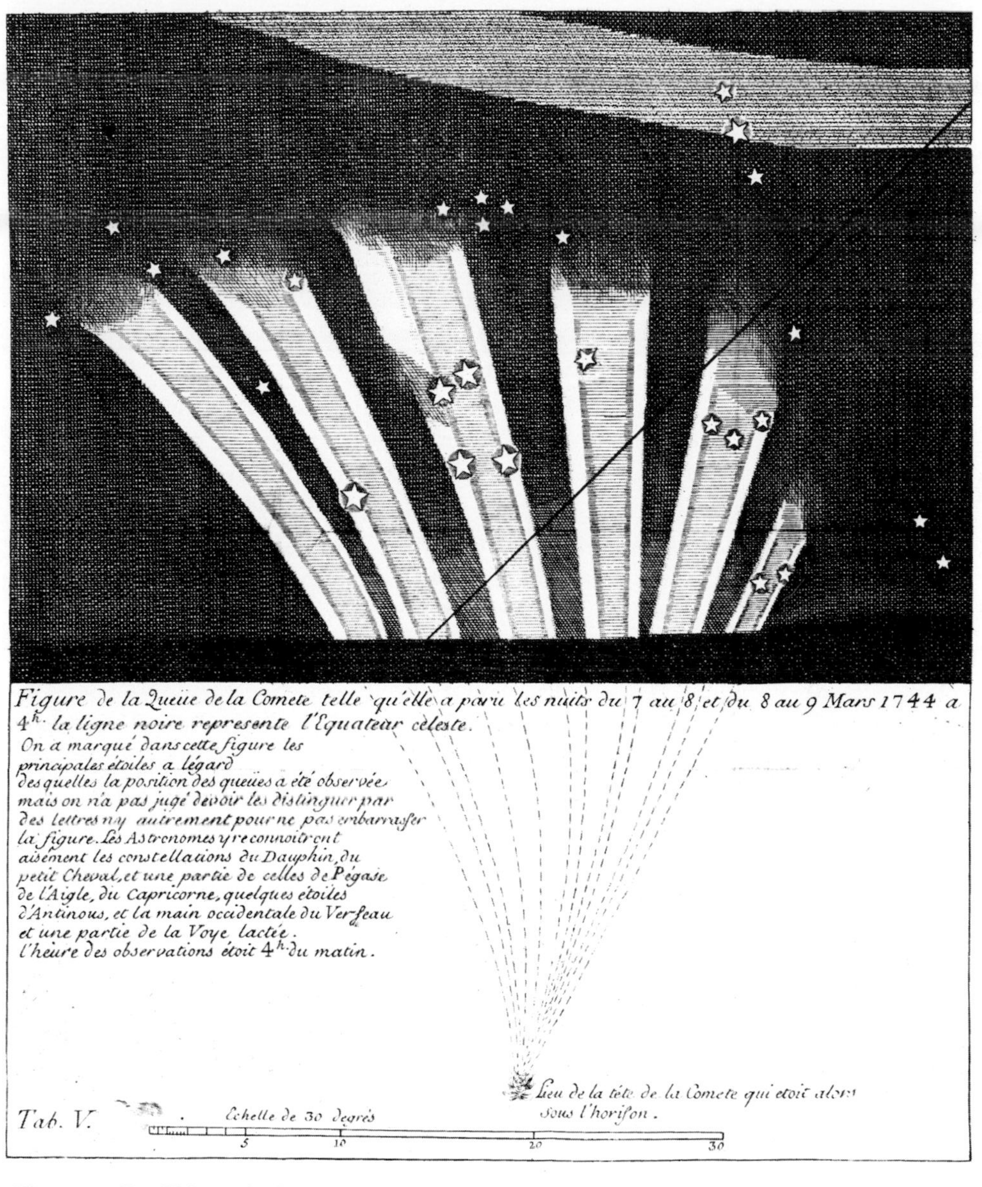

Fig. 61. *De Chéseaux's Comet of 1744*, from J. P. Loys de Chéseaux, *Traité de la comète . . .*, 1744.

the French astronomer Joseph Jérôme le français de Lalande, who published a corrected version of Edmund Halley's tables, and who was best known as a popularizer of astronomy, delivered a lecture entitled "Reflections on Those Comets Which Can Approach the Earth," in whose wake a rumor spread in Paris that on May 20 a comet would destroy the earth.

The growing observational obsession is reflected in a series of eight paintings entitled *The Astronomical Observations* (dated 1711, in the Vatican Collection) by the Italian artist Donato Creti. It features astronomers with their telescopes, Saturn and its rings, the moons of Jupiter, and a comet in the last scene, where a woman sits in front of foliage at the left. She is turned to face the viewer as though arrested in the action of placing a nosegay of flowers in her hair and is unaware of the comet plunging downward in the nocturnal sky. (Her petite, delicate proportions, like those of the entire series, are characteristic of rococo refinement.) In the middleground directly behind her, a pastoral couple sits on the bank of a river, again unaware of the celestial drama, while on the right bank two standing figures regard the blazing comet. This comet was most likely based on an astronomer's sketch of one of the early naked-eye apparitions of the eighteenth century, either in 1702 or 1707.

The series has a most fascinating history worth recounting here because it is directly linked to telescopic observation. Creti's patron, Count Luigi Ferdinando Marsili, commissioned the series of eight paintings in 1711 to present to Pope Clement XI, hoping to obtain the Pope's support for construction of an observatory in Bologna. In 1703, Marsili had had an observatory built in his own palace and entrusted his precious new equipment to an astronomer, Eustachio Manfredi. Creti followed Manfredi's instructions and also incorporated some of the Count's astronomical equipment in the paintings. A miniaturist, Raimondo Manzini, added the celestial bodies, which, except for the comet, are depicted as viewed through a telescope, explaining their abnormal dimensions in proportion to the Venetian-inspired landscapes. The paintings, which have been in the Papal Collection since 1711, were convincing, because Marsili obtained the Pope's support to build both an institute of science (where Manfredi became the director) and an observatory.

In 1744 a very odd comet appeared. It inspired a poem by Mather Byles, whose frontispiece depicts the first comet represented in the New World hovering over Boston, where it is telescopically observed by colonial Bostonians (Fig. 60). The comet was discovered by a Dutchman on 9 December 1743, then sighted four days later by the Swiss astronomer J. P. Loys de Chéseaux; it is sometimes therefore referred to as De Chéseaux's Comet. A few days after its perihelion in March 1744, a multiple, peacocklike tail was observed (Fig. 61), which at one point manifested eleven fanning rays. Quite understandably, De Chéseaux's Comet attracted a great deal of attention, although it remained brilliant for only a short time.

This comet may have even inspired the comet motif included by William Hogarth in the first print of his scathing indictment of society, *Marriage à la Mode* (Fig. 63). The prints, based on a series of his paintings, were published in 1745; they satirize the nouveau riche, the degenerate aristocracy, and the English mania for the French style. The first scene depicts the signing of a marriage contract between Lord Squanderfield (on behalf of his sniveling syphilitic son) and a newly rich merchant (on behalf of his weak, flirtatious daughter). Hogarth included a comet in the pretentious portrait of Lord

Fig. 62. Thomas Rowlandson, *The Comet*, dated 1821 in another hand, ink and watercolor over pencil drawing, 13 x 10⅝ in. Formerly, New York, Christie's.

Fig. 63. William Hogarth, detail of *Marriage à la Mode*, plate I, 1745, engraving.

Squanderfield, which is shown as painted in the French style, then all the rage in London. In this portrait, Squanderfield stands in an arrogant pose and wears the insignia of the Order of the Golden Fleece. He is flanked by a cannon, which fires directly at the doomed young couple, while an ill wind blows and a comet hovers overhead. Nothing in Hogarth's composition is accidental, so that every detail must be read as a pointed allusion. The comet here is both an omen of the approaching ruin of the frivolous couple and a satirical comment on Squanderfield's pomposity.

Another ominous comet depiction of the era, with a novel sardonic twist, appears in Johann Heinrich Fuseli's print *Fortune Suspended Over the Earth* (Fig. 64). This traditional subject receives a more modern interpretation as a tipsy, slightly menacing Fortune dominates the composition. She is dressed as a jester to make her appear more like fickle Chance. A disembodied comet careens in from the right to add another pernicious and unstable chord.

Certain artistic conventions were employed in the rendering of comets and these carried over to images of other celestial events. For example, the English artist Thomas Sandby recorded a large meteor observed in August 1783 (Fig. 65) which was identified with the ancient type called *Draco volans*, or

Fig. 64. Johann Heinrich Fuseli, *Fortune Suspended Over the Earth*, 1780–90, etching.

"flying dragon" (Fig. 66). Although Sandby's watercolor at first appears to be a realistic observation (Fig. 65), a second look reveals the work's hauntingly surreal quality, which results both from the deserted stillness of the unfinished work and from the artist's strange depiction of the meteor traveling across the sky three times, as it develops a tail. The effect is that of three superimposed still photographs. The watercolor is from an album whose

Fig. 65. Thomas Sandby, *The Meteor of 18 August 1783 in Three Aspects Seen From the Northeast Corner of the Terrace, Windsor Castle*, watercolor, 11¼ x 18 in. London, The British Museum.

Fig. 66. Henry Robinson, *Meteor Seen on 18 August 1783*, etching.

subsequent page contains a similar design of the same scene without the sky and landscape but with a group of figures watching from the terrace. A third watercolor in the Royal Collection combines both compositions. The entire scene was also recorded in an aquatint by his brother Paul Sandby. It is known that Thomas Sandby viewed this meteor with other persons, including Tiberius Cavallo, the Italian-born physicist, who published an account of the phenomenon in the Royal Society's *Philosophical Transactions* in 1784. The meteor's peculiar horizontal motion and "sputtering" tail seem to be eighteenth-century conventions used for comets, as seen in Scott's painting of Halley's Comet in 1759 (Fig. 19).

Thomas Rowlandson, the English caricaturist, drew an amusingly bawdy illustration (Fig. 62) of comet-struck individuals gazing at a rare daylight comet; telescopes are here shown in the possession of amateurs. One enterprising rake takes advantage of the comet's mesmerizing power to make amorous advances to his callipygous paramour. This drawing, which portrays virulent comet fever, is quintessentially eighteenth-century English in sentiment, although it dates from the early years of the nineteenth century, the comet-crazy century, when Rowlandson rendered a group of comet satires, one of which is entitled *Looking at the Comet Till You Get a Criek in the Neck*.

Five

The Comet-Crazy Century

> "O tell me, Sir Comet," great Greeley [Horace] said,
> "Why so large a tail to so small a head?"
> "Why sir," replied the Comet, "my head, 'tis true,
> Unlike yours, looks small, but like yours loves to
> Spin out a fabulous tale."
>
> —*Vanity Fair*, 1861

Because of its many spectacular apparitions (about thirteen naked-eye comets and a total of three hundred comets sighted all together), the nineteenth century should rightfully be known as "The Comet Century." These comets passed over the earth during a pivotal period in Western civilization, one that experienced the gigantic upheavals of the Industrial Revolution, Nationalism, Positivism, the theories of Freud and Marx—in short, nothing less than the birth of the modern age. In 1882, the first generally acknowledged good scientific photograph of a comet (Fig. 67) was taken, changing the course of cometology forever.

While the people of this century were basically motivated by an intense romantic longing, scientists put cometology on a more objective basis. They studied comets as part of celestial mechanics, resulting in a more thorough understanding of the physical nature of comets. Around the turn of the century, the distinct natures of meteors, meteorites, and asteroids were more firmly defined, but it was not until midcentury that meteors and comets were clearly distinguished, and the axiom "If it moves and is in the heavens, call it a comet" no longer applied.

Throughout the century, as the understanding of optics grew more sophisticated, larger refracting telescopes were constructed, which substantially aided celestial sleuthing. One compact type, named The Comet-

Watcher, was especially popular for charting cometological data. Consequently, much of the remaining mystery and lore surrounding comets—that had not already been dispelled by the momentous discoveries of the seventeenth and eighteenth centuries—disappeared for natural scientists during the century. Because of the long, often fruitless hours that serious comet hunting requires, comet observation during this century was as much a practice of amateurs as professionals.

But the public remained largely unaffected by the new understanding and preferred either to cling to age-old superstitions of comets as instant doomsday machines (Fig. 68), or to invent contemporary fantasies based on old themes. This condition was only natural, since the era was, in general, prone to excess and hysteria, both of which tended to exacerbate recurring bouts of virulent comet fever. Indeed, the nineteenth century maintained a dualistic attitude toward comets, as exemplified by Thomas Coleridge, the great Romantic poet. He looked at nature in a mystical manner while attempting to understand comets in a scientific fashion. He failed in the latter, however, as shown in a letter of 1820 in the British Museum in which Coleridge argues against Newton's theory.

As the century opened, the diminutive Napoleon Bonaparte cast his large shadow over European civilization. Never a man to doubt his own powers, and a master of self-promotion, he cannily adopted various comets as his protecting *genii*. He was no doubt well aware of the tradition wherein comets were associated with kings and great rulers. By quickly adopting comet symbolism, Napoleon gave his reign a traditional resonance and legitimacy. The first historical comet linked with Napoleon was the Great Comet of 1769 (often called Napoleon's Comet), which by all reports had an unusual red luster and a tail spanning more than sixty degrees. Since portents can be interpreted in various ways for propagandistic purposes, it was later interpreted by his enemies as foreshadowing bloodshed, destruction, and devastating war; and by his supporters as a triumphant sign of his glorious reign.

When the Great Comet of 1811–12, a brilliant daylight comet also referred to as Napoleon's Comet, appeared, Bonaparte enthusiastically greeted it as his guiding star and the controller of his destiny. What fun an English caricaturist made out of the situation by personifying Napoleon as a malevolent comet covetously eyeing England and inscribing his print *A Corsican Comet/Frenchified*! As he assembled the greatest army in Europe since Xerxes had invaded Persia, Napoleon claimed that this comet foretold his victory in the Russian Campaign. Ironically, the reverse proved to be true, for while the comet blazed in the sky, Napoleon was well on his way to being utterly defeated by the cruel Russian winter.

The Great Comet of 1811–12 was not only the most spectacular naked-eye comet of Bonaparte's reign, but also ranks as one of the most ominous com-

Fig. 67. Sir David Gill, *Photograph of the Great September Comet of 1882*.

Fig. 68. Honoré Daumier, *Ah! . . . comets . . . they announce all kinds of bad things . . . poor Madame Galuchet died suddenly this night . . .*, 1858, lithograph.

ets of modern times. Since it was visible for seventeen months, it holds the record for the longest naked-eye apparition in recorded history. The huge comet's tail reached a span as wide as seventy degrees in December 1811, and it seems to have displayed the largest coma ever recorded. Its maximum diameter is estimated to have been 1,250,000 miles, considerably larger than the sun's, yet with only a tiny fraction of the sun's mass. As Napoleon and the *grande armée*, some seven-hundred-thousand strong, marched into Russian territory, the comet burned with an intensely brilliant light, and its tail performed frightening acrobatics by splitting into two. Around October, the tail reached a maximum length of about one hundred million miles and a breadth of fifty million miles. None of us will probably ever see its like again unless another rogue comet comes into view, since its period has been estimated—not without some controversy—at about 3,065 years, give or take 50. No wonder, then, that a world working its way out of the Enlightenment's bog of rationality regarded the stupendous Comet of 1811 with awe. Thomas Hardy was duly mesmerized by its apparition; in his novel *Two on a Tower*, the tower of the title being an improvised observatory, his astronomer protagonist estimates its period at thirty centuries.

As Napoleon's troops were taking Moscow in September 1812, the Great Comet was no longer visible to the naked eye, but between July 20 and September 21, the Pons-Brooks Comet made its first recorded apparition. Also plainly visible for a brief time, it was confused with the earlier comet by nonscientists. It is therefore logical that Leo Tolstoy exercised artistic license and used the fiery comet of 1811–12 in *War and Peace* when, as Napoleon's troops invade the Russian capital, the Great Comet functions not only as a sign of war but as a transcendental metaphor for the renewing power of love:

> It was only looking at the sky that Pierre forgot the mortifying meanness of all things earthly in comparison with the height his soul had arisen to . . . the immense expanse of dark starlet sky lay open. . . . Almost in the centre . . . , surrounded on all sides by stars, but distinguished from all by its nearness to the earth, its white light and long, upturned tail, shone the huge, brilliant comet of 1812; the comet which betokened, it was said, all manner of horrors and the end of the world. But in Pierre's heart that bright comet, with its long luminous tail, aroused no feeling of dread. On the contrary, his eyes wet with tears, Pierre looked joyously at this bright comet, which seemed as though after flying with inconceivable swiftness through infinite space in a parabola, it had suddenly like an arrow piercing the earth, stuck fast at one chosen spot in the black sky, and stayed there vigorously tossing up its tail, shining and playing with its white light among the countless other twinkling stars. It seemed to Pierre that it was in full harmony with what was in his softened and emboldened heart, that had gained vigor to blossom into a new life.

Fig. 69. *The Effects of the Comet of 1811*, engraving. For years, the comet wines of 1811 were coveted and featured proudly in price lists and advertisements.

Although Napoleon claimed the Great Comet of 1811 as his own, the *hoi poloi* reacted with the traditional fear of such celestial events. This is shown in a French print of the time (Fig. 69), in which the comet is personified as a woman, here threateningly brandishing two flaming torches. She is literally a "hairy star," for her flowing locks form a bifurcated tail of fire, which deals out death and destruction—even a volcanic eruption to remind us that the 1759 eruption of Vesuvius was still a vivid memory. The print is also an illustration of an old German rhyme:

> Eight things there be a comet brings
> When it on high doth horrid rage:
> Wind, Famine, Plague, and Death of Kings
> War, Earthquake, Floods, and Doleful Change.

Fig. 70. Luigi Calamatta, *Still Life With Death Mask of Napoleon*, 1834, engraving.

Fig. 71. Charles Bouvier, after Carl von Steuben, *The Eight Epochs of Napoleon*, 1837–42, engraving.

In 1802 long before the spectacular apparition of 1811, Napoleon's comet symbolism, together with his imperialistic goals, appear in a bizarre and mannered painting by Anne Louis Girodet-Trioson, *Ossian Receiving the Napoleonic Officers* (Fig. 72). It is based on a scandalously fake poem, supposedly by the Gaelic bard Ossian, but actually by the enterprising hoaxer William MacPherson, that was the rage of all Europe as well as the Emperor's favorite. Girodet's painting depicts a group of deceased French heroes, who are introduced by the floating goddess of victory into a foggy, Nordic Valhalla, where they are welcomed by the blind bard Ossian and his ghostly entourage. These ectoplasmic shades are painted in a style that was influenced by the light experiments of Antoine-Laurent Lavoisier, as well as by contemporary theatrical stage effects using "magic lanterns" to send phosphorescent ghosts gliding across the stage. Such light tricks delighted and titillated the occult-mad Parisians. The identity of the lights in the painting's upper right-hand section have been extensively debated; are they meteors, northern lights,

Fig. 72. Anne Louis Girodet-Trioson, *Ossian Receiving the Napoleonic Officers*, 1802, oil, 36½ x 72½ in. Musée National du Château de Malmaison.

or comets? One is clearly a comet; logically enough, since Napoleon associated comets with victories, it is poised directly above Victory herself. Other lights may be tailless comets, many of which appeared during the early years of the century. Still others may be meteors, for Bonaparte is said to have remarked that geniuses are meteors destined to burn up while illuminating their centuries. Because the painting's complex meaning concerns the glorification of French heroes who died for their country—and hence, by implication; for their Emperor—Girodet's operatic fantasy is a veiled celebration of Napoleon's role as victor and peacemaker during 1801–02. In addition, in an innovative nineteenth-century manner, Girodet inverted the traditional pejorative association of comets with revolution and war to invent a positive symbolism wherein they represent the triumphant revolutionary mood of 1801.

Napoleon's entire career is satirized in a print whose message is that the hat *is* the man (Fig. 71). A comet is ironically poised above the vignette of Napoleon's debacle, the Battle of Waterloo in 1815. Since a comet appeared at Napoleon's birth, another was supposedly spotted over St. Helena and France one night before his death in 1821, in order to fit the tradition associated with kings and rulers. A comet also accompanied Napoleon after his death, as seen in a print depicting his death mask and astrological symbols, where it hovers over the once-imperial forehead (Fig. 70). This tradition was imitated during the reign of Napoleon III (1852–70), as seen in a painting by Victor-Jean Adam in a private collection that shows a comet in conjunction with his apotheosis.

English artists also capitalized on Napoleon's Romantic infatuation with comets; for example, Rowlandson's 1807 print *John Bull Making Observations on the Comet* (Fig. 73) uses Napoleon's profile as the comet's head. Rowlandson must have possessed a smattering of scientific knowledge, because he accurately represented the comet as reflecting the sun in his clever rendering; and the sun is none other than George III. Napoleon's ill-starred heir did not escape his dynastic inheritance, as seen in a print of 1811 (Fig. 74) in which he is personified as the comet that leads the Magi to the Holy Family—none other than Napoleon, Marie-Louise, and again their son, the infant King of Rome. How heretically far have the Magi journeyed since Giotto's *Adoration of the Magi* (Fig. 15)!

In the nineteenth century, comets became more than mere vagabonds of the solar system, to be watched and studied; they became popular symbols that haunted the imagination, that universe more real to many Romantics than physical reality. Artists and writers wishing to escape sordid reality often represented comets in their work that did not refer to specific apparitions. Sometimes these nonspecific comets were part of an already existing literary and/or artistic tradition and thus symptomatic of the age's historicism. The Italian illustrator Bartolomco Pinelli's 1826 representation of the following passage from Dante's *Paradiso* is a case in point:

> Thus, Beatrice; and those souls beatified
> Transformed themselves to spheres on steadfast poles
> Flaming intensely in the guise of Comets.
>
> (Canto XXIV, Longfellow translation)

Fig. 73. Thomas Rowlandson, *John Bull Making Observations on the Comet*, 1807, colored engraving.

Fig. 74. William Elmes, *The Gallic Magi Led by The Imperial Comet*, 1811, colored engraving.

Romantic rhetoric frequently personified intensely passionate individuals as comets. Lord Byron wrote of Charles Churchill that he "blazed as a comet of a season," while William Blake, both poet and painter, has personified the comet of Edward Young's "Night Thoughts, 149" as a plunging androgynous being in his illustration (Fig. 75) of the following lines:

> Has thou ne'er seen the Comet's flaming Flight.
> The illustrious stranger passing; terror sheds
> On gazing Nations from his fiery Train. . . .
> Thus, at the destined Period shall return
> He, once on Earth, who bids the Comet blaze
> And with him all our Triumph O'er the Tomb.

In "Night Thoughts," Young refers five times to comets—one time as a symbol of the life of the soul, where one tragic individual plummets terrifyingly toward annihilation, and another as a positive omen: "Comets good omens are, when duly scann'd. . . ."

Blake, himself a mystic of great complexity, was totally captivated by comet symbolism. For example, in his frontispiece illustration to Thomas Gray's *Poems* of 1797, Blake portrays a comet as a metaphor of ascendant Fame—swift, brilliant, and, alas, perhaps fleeting. He also mentions and depicts comets or shooting stars four times in *Milton: A Poem in Two Books*, again a marriage of poetry and the visual arts (Fig. 76). Each time a comet appears in *Milton*, it symbolizes poetic inspiration and embodies the soul of the great English Puritan poet who also used many comet references, especially in *Paradise Lost* and a *History of Britain*. In Figure 76, the bard Milton, represented in the form of a comet or shooting star, calls out Blake's name, "WILLIAM," as it charges like an electric current through the artist's spasmodically racked body. Blake also associated comets with rulers in his arcane political allegory *The Spiritual Form of Pitt Guiding Behemoth;* this apocalyptic scenario is based on Blake's "vision," which according to the artist loomed over one hundred feet high.

Even more bizarre is his hallucinatory painting *The Ghost of a Flea* (Fig. 77), which contains the artist's most memorably rendered comet. John Varley, an artist friend who was interested in astrology, recorded that Blake described his own vision as follows: "Here he comes! his eagre tongue whisking out of his mouth, a cup in his hands to hold blood, and covered with a scaley skin of gold and green. . . ." Varley also relates that while Blake drew the preparatory sketch for the painting, "the flea told him that all fleas were inhabited by the souls of such men, as were by nature bloodthirsty to excess." The comet (incidentally, comets are also referred to as "bloodthirsty") in Blake's painting functions as a symbol alluding to the blood and destruction caused by the rapacious nature embodied in the flea. How insidiously terrifying Blake's vision is even today! His image of human wickedness, a specter half-human and half-beast, dances in an incongruous posture, his bloodthirsty tongue flicking like that of a lizard and his bulging eyes aglow with infernal

(40)

Of Pomp, and Multitude; a radiant Band
Of Angels new; of Angels from the Tomb.

Is this by Fancy thrown remote? and rife
Dark Doubts between the Promife, and Event?
I fend thee not to Volumes for thy Cure;
Read Nature; Nature is a Friend to Truth;
Nature is Chriftian, preaches to Mankind;
And bids dead matter aid us in our Creed.
Haft thou ne'er feen the Comet's flaming Flight?
Th' illuftrious Stranger paffing, Terror fheds
On gazing Nations, from his fiery Train
Of length enormous; takes his ample Round
Thro' Depths of Ether; coafts unnumber'd Worlds,
Of more than folar Glory; doubles wide
Heaven's mighty Cape, and then revifits Earth,
From the long Travel of a thoufand Years.
Thus, at the deftin'd Period, fhall return
He, once on Earth, who bids the Comet blaze;
And with Him all our Triumph o'er the Tomb.

Nature

Fig. 75. William Blake, Illustration to Young's "Night Thoughts, 149," 1795, watercolor, 16⅛ x 12⅝ in. London, The British Museum.

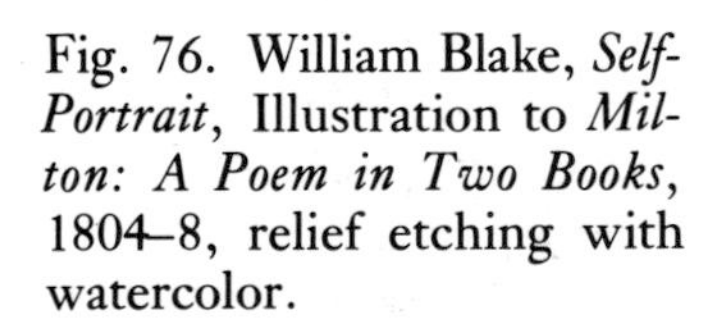

Fig. 76. William Blake, *Self-Portrait*, Illustration to *Milton: A Poem in Two Books*, 1804–8, relief etching with watercolor.

Fig. 77. William Blake, *The Ghost of a Flea*, 1819–20, tempera heightened with gold on panel, 8½ x 6⅜ in. London, The Tate Gallery.

Fig. 78. Jean-François Millet, *The Shooting Stars*, 1847–48(?), oil, 7 x 13 in. Cardiff, National Museum of Wales. Here, two couples, each a personification of a shooting star, are wound inextricably together in erotic embraces as they float through the starry sky, rather like Paolo and Francesca in Dante's second circle of Hell in *The Inferno*.

light. Between the ghost's feet rests the tiny flea. The comet with its phosphorescent glow is most convincing, for its coma and tail convey the vaporous reflection and fluorescence of actual comets. Most likely, the image reflects Blake's memory of either the Great Comet of 1811 or the bright Comet of 1819.

Certainly the Comet of 1811 and Halley's 1835 apparition must have inspired the painter John Martin in his operatic composition *The Eve of the Deluge* (Fig. 79). (Although four comets appeared in 1840 and the first one was visible with the naked eye, none was especially memorable.) Martin, a believer in natural events as signs or judgments, specialized in sweeping apocalyptic canvases, which later influenced the filmmaker D. W. Griffith, and perhaps Francis Ford Coppola's sets for *Apocalypse Now*. Martin's Genesis scene is echoed by the following lines from Byron's "Heaven and Earth":

> The earth's grown wicked
> And many signs and portents have proclaimed
> A change at hand, and an overwhelming doom
> To perishable beings.

Martin's stupendous comet plunges downward toward the mist-enshrouded sun and rising moon, a vivid prophecy of the impending flood. At the lower right, the Israelites continue their revels, heeding neither the comet nor the ravens (another ill omen), while on the craggy promontory the family of Noah gathers around Methuselah, who consults the scrolls of his father, Enoch, and attempts to read the portents. Immanuel Velikovsky, the twentieth-century cosmologist, among others, has actually credited the biblical flood to a comet!

Comets and meteors, still used interchangeably, were also employed apocalyptically in contemporary literature, as Nathaniel Hawthorne did, for example, in *The Scarlet Letter*:

> "At the great judgment day," whispered the minister. . . . But before Mr. Dimmesdale had done speaking, a light gleamed far and wide over all the muffled sky. It was doubtless caused by one of these meteors [comets]. . . .They stood in the noon of that strange and solemn splendor, as if it were the light that is to reveal all secrets, and the daybreak that shall unite all who belong to one another. . . . Nothing was more common in those days than to interpret all meteoric appearances, and other natural phenomena . . . as so many revelations from a supernatural source. Thus, a blazing spear, a sword of flame. . . . It was, indeed, a majestic idea, that the destiny of nations should be revealed, in these awful hieroglyphics, on the cope of heaven. . . .We impute it, therefore, solely to the disease in his own eye and heart, that the minister, looking upward to the zenith, beheld there the appearance of an immense letter,—the letter A. . . . "But did your reverence hear of the portent that was seen last night?—a great red letter in the sky—the letter A, which we interpret to stand for Angel. For, as our Good Governor Winthrop was made an angel this past night. . . ."

During the nineteenth century, eroticism and comets could go hand in hand. Comets and their cousins, meteors or shooting stars, were discreet allusions to adultery (*The Scarlet Letter*) and to transcendental love (*War and Peace*), and even simply to the very act of physical love, as in the socialist Jean-François Millet's painting *The Shooting Stars* (Fig. 78). Millet's figures themselves emit light, and the red contours around the man of the left-hand couple communicate the heat of passion, while the elongation of the figures formally conveys their longing voluptuousness, transient as shooting stars.

The German artist Carl Gustav Carus—who was also a physicist, physician, and philosopher—depicted a highly symbolic comet in a drawing executed about 1851 (Fig. 80). It is difficult to determine if the artist was influenced by a specific apparition, for in the early 1850s several comets appeared, two in 1850, four in 1851, and four in 1852, the third of which was the last apparition of Biela's Comet. The window motif in Carus's drawing—a favorite device of Romantic artists to demonstrate their attitude toward nature—sets a meditative, melancholy mood; the Gothic cathedral adds the requisite craggy, picturesque silhouette; and the comet lends sublimity. Its eerie light dominates the old symbol of the Christian God, the cathedral, in "God's playground," the sky, so that we the viewers feel powerless before the magnificence and capriciousness of nature. The telescope and globe in-

Fig. 79. John Martin, *The Eve of the Deluge*, 1840, oil, 56½ x 85½ in. The Collection of H.M. The Queen. (For detail see frontispiece.)

dicate Carus's scientific interests. The mood of his drawing parallels the following passage by Gustave Flaubert from *Madame Bovary*:

> Later with Sir Walter Scott, she [Emma Bovary] developed a passion for things historical. . . . She wished she could have lived in some old manor house, like those chatelaines . . . who spent their days with their elbows on the stone sill of a Gothic window surmounted by a trefoil. . . . In those days, she worshipped Mary Queen of Scots and venerated other illustrious or ill-starred women. For her, Joan of Arc, Héloise, Agnès Sorel, La Belle Ferronière and Clémence Isaure stood out like comets against the dark vastness of history.

A similar mystical attitude is found in a much more realistic rendering of a meteor shower, closely resembling the summer Perseids, in Millet's nocturne *Nuit Etoilée* (ca. 1850–57, in the Yale University Art Gallery, New Haven), set in the Forest of Fontainebleau just after sunset. While the painting appears to be an intensely realistic observation of landscape and the meteorological effects of the night sky, a passage from one of Millet's letters reveals his Romantic, mystical attitude toward nature: "If only you knew how beautiful the night is! There are times when I hurry out of doors at nightfall . . . and I always come in overwhelmed. The calm and the grandeur of it are so awesome that I find I actually feel afraid."

For Millet and others, meteors and comets still signified a loss of order in the universe. Millet's wonder and quiet fear is reinterpreted at fever pitch in

Fig. 80. Carl Gustav Carus, *View Through a Window With a Comet*, ca. 1851, charcoal drawing, 17¼ x 10¾ in. Dresden, The Kupferstichkabinett.

Fig. 81. *The Great Comet of 1843 Seen in Paris on 19 March 1843*, lithograph. This splendid daylight comet, also a sungrazer, dominated the sky as seen in this realistic print, a type used to record cometological phenomena before photography.

Vincent van Gogh's hallucinatory canvas *The Starry Night* (Fig. 83), admittedly influenced by Millet's work and painted while the artist was a resident in the mental asylum of Saint-Rémy. Van Gogh's pulsating forms of throbbing light, some resembling identifiable heavenly bodies including a spiral nebula and a comet, swirl like wild dervishes in the cosmic mix of the heavens.

The Great Comet of 1843 (Fig. 81) inspired Honoré Daumier, the gifted French illustrator and caricaturist, to depict a large comet suspended over the melancholy Victor Hugo (Fig. 82). The print's amusing, rhymed caption asks why the tail of the comet is longer than the ticket lines for Hugo's new and unsuccessful play *Burgraves.* Here the comet functions as an ironic metaphor of fame. Daumier's satirical, detached tone about comets' portentous nature is reminiscent of Alexandre Dumas's remark in *The Count of Monte Cristo,* "What's the matter, baron? . . . You look disturbed and that frightens me, a worried capitalist is like a comet, he always presages some disaster for the world."

Comets are always newsworthy, but by 1843 even the press could be detached about comets, as revealed in the following announcement of the Comet of 1843 published in the *London Illustrated News*: "Thereby hangs a tail.—Shakespeare." The "tail" refers to the comet's anatomy, a wordplay on the famous Shakespearian line ending with "tale."

The brilliant daylight Comet of 1843 was superior to the Comet of 1811, according to one astronomer who saw both. It boasted the longest recorded tail, an estimated two hundred million miles in length—greater than the distance between the sun and Mars. According to an astronomer in Naples, its tail was bright enough to be noticeable above Vesuvius during a full moon. Besides, it was a sungrazer, passing at perihelion within one hundred thousand miles of the sun's surface.

George Cruikshank's print *Passing Events on the Tail of the Comet* (Fig. 84), published on January 1, 1854, and its preparatory watercolor in the Victoria and Albert Museum, London, reflect the public's consciousness of comets. The works both feature a grinning comet whose tail encompasses all noteworthy and trivial events of that entire year, drawn in antlike scale. While there were four comets observed in 1853, two of which were visible with the naked eye, and one comet observed in January of 1854, Cruikshank certainly did not intend to refer to a specific comet.

Fig. 82. Honoré Daumier, *Hugo Looks at His Depression*, 1843, lithograph.

Fig. 83. Vincent van Gogh, *The Starry Night*, 1889, oil, 29 x 36¼ in. New York, Collection, The Museum of Modern Art, acquired through the Lillie P. Bliss Bequest.

Comets remained a topical issue throughout the century—and no wonder, for in 1857 alone, seven comets were recorded. By midcentury, astronomy, which for the masses meant comet-seeking, was a popular fad and thus fodder for contemporary satire. In *A Surprise*, by Daumier (Fig. 87), an amateur astronomer stands before his telescope. He has blindly failed to find his obvious quarry, causing the comet to tap the startled man on the shoulder to correct his myopia. In another Daumier print, the aspiring astronomer Babinet appears to be intensely focused on a comet but, as is pointed out, he too has missed it altogether. Is Babinet really that inept, or is he merely a voyeur, using his telescope to look at other "celestial bodies" in his neighbors' windows? In 1857, the year of seven comet apparitions, Daumier was still afflicted by the comet bug and satirized the contemporary bourgeoisie's gullible attitude toward comets in a series of ten cartoons for the magazine *Charivari*. He continued this theme with nine additional prints in 1857, and with two lithographs in 1858, a year of eight comet apparitions. In fact, Daumier holds the record for creating the largest number of comet representations. The universality of his insight into human nature is painfully evident in one work (Fig. 88) in which a street thief picks the pocket of an unsuspecting gentleman, who gazes at the sky, transfixed by the comet. The costumes in this print are nineteenth-century, but the situation is all too modern! In 1857, as cometomania raged, a cartoon depicting a "hairy star" viciously tearing the earth asunder (Fig. 7) was published at the end of the apparition of the short-period Brorsen's Comet, which was observed from March 18 until June 22. This bright comet has been seen only five times, the last in 1879. Despite our smugness and sophistication, comet fear still endures; we have only to remember the mistakenly named television movie *Meteor*, whose plot revolved around a comet colliding with the earth, or the 1984 movie *Night of the Comet*.

Another type of fear regarding comets is satirized in a Daumier print (Fig. 89) lampooning not only the German obsession with comets but also the French periodical *Le Canard*, famed for its prognostications. The print makes a complex pun on the word *canard*, which is French for both "duck" and "false news" or "hoax." In the print, a mad astronomer is dressed in the same charlatan's costume as featured in the paper's logo.

Then, in 1858, Donati's Comet (Figs. 90 and 92), which is considered by many to be the most beautiful comet in history, made its first and only recorded appearance. On June 21 in Florence, Giovanni Battista Donati first sighted his famous namesake telescopically. By August 29, Donati's Comet was visible to the naked eye, and it remained observable by telescope until March 4, 1859. Its fame derives from its wonderfully unusual profile, characterized by one large dust tail and two thin plasma or ion tails. The interaction between the two types produced a symphony of curves and knots in the formation of all three tails. Donati's Comet also had a number of fluctuating envelopes of light (three around the head and seven around the entire comet), which were recorded by astronomers in delicately shaded drawings. The detailed information in these scientific drawings is astounding and is still invaluable to cometologists today. For example, in 1978, Whipple retroactively determined the rotation rate for Donati's Comet by deducing from these drawings the intervals of the dust emissions that formed the halos. These emissions arose from an especially active area on the surface of the comet's nucleus stimulated during its rotation into sunlight.

No one knows definitely whether Donati's Comet will ever return to give us a chance to photograph it, although cometologists anticipate that it is a periodic comet positioned in an elliptical orbit in which it returns to perihelion roughly every 2,000 years. One mathematician has gone so far as to

Fig. 84. George Cruikshank, *Passing Events on The Tail of the Comet of 1853*, from *George Cruikshank's Magazine*, January 1, 1854, etching. In the humorous concept of this image, the comet is a clever device to sum up the old year. An enthroned Queen Victoria and her consort, Albert, preside over a potpourri of contemporary scandals, temperance rallies, peace conferences, the war between Russia and Turkey, Guy Fawkes Day, Derby Day, and other Victorian events.

postulate that Donati's Comet may be identical with the comet recorded by Seneca in 146 B.C.

On October 5, 1858, five days before perihelion, Donati's Comet passed in front of the very bright star Arcturus, to the left of the comet shown in Figure 92 and to the right of the comet shown in Figure 90. To the surprise of everyone, neither dimmed the light of the other. It is estimated that at this

time the comet's tail (spanning around forty degrees) had grown, from a length of fourteen million miles in August, to fifty million miles.

So wondrous were the pyrotechnics of Donati's Comet on October 5 that the periodical *The Annual Reporter* declared, "The population of all the western world was probably out-of-doors gazing at the phenomena," while the *London Times* remarked that while the solar system becomes so ordinary that even ladies' maids understand it as easily as needlework, the spectacle of a comet remains awe-inspiring.

Several artists were among the dedicated watchers of Donati's Comet, but their seemingly realistic renderings of the apparition were often highly manipulated. For example, William Turner of Oxford's arresting watercolor (Fig. 85) is not the accurate recording of the sensational night of October 5 that its pseudo-scientific title would have us believe, for it includes neither

Fig. 85. William Turner of Oxford, *Donati's Comet, Oxford, 7:30* P.M. 5 October 1858, watercolor, 10⅛ x 14⅜ in. New Haven, Yale Center for British Art, Paul Mellon Collection.

the star Arcturus nor the characteristic ion tails. Donati's Comet is also represented in a second seemingly realistic work, *Pegwell Bay, Kent, October 5, 1858* (Fig. 86)—by another English artist, William Dyce—in which it is shown at sunset, one of the times when all comets except those rare daylight comets first become visible. The almost photographic realism of the painting not only reflects Dyce's Pre-Raphaelitism but also demonstrates the artist's familiarity with the photography of David Octavius Hill. Upon closer analysis, however, the painting is more than just an objective rendering of the comet's apparition over the coastline of Kent (where the artist's family was on holiday); it is a rather complex allegory of the passing of time and man's vulnerability in the universe. In short, it is a Victorian confrontation between the exciting discoveries of science and the persistence of old superstitions. In the context of Dyce's painting, Donati's Comet symbolizes the longevity of the forces of the universe and nature (here, the tide), which dwarf the human figures and eventually wear them down (like the cliffs of the Kent coastline). Moreover, the mood of impending doom is heightened by the sense here that Donati's Comet will become increasingly brighter and more fearsome as night sets in.

During the mid-nineteenth century, several other memorable comets careened across the sky, two of which were commonly associated by Americans with the Civil War. In 1861, Comet Tebbutt appeared, first spotted by the amateur astronomer J. Tebbutt in Windsor, Australia, on May 13. According to contemporary descriptions, while this naked-eye apparition did not surpass Donati's Comet in beauty, it had other unique attractions. For example, it was decidedly more brilliant and had a yellowish-red head, a complex nucleus, eleven surrounding envelopes, and an extremely long tail estimated at between 90 and 120 degrees. It was so bright between June 29 and July 1 that it cast shadows at night! Tebbutt's Comet was also visible during daylight and is thought by some to have been the brightest comet of the century.

Fig. 86. William Dyce, *Pegwell Bay, Kent, October 5, 1858*, oil, 25 x 35 in. London, The Tate Gallery.

Fig. 87. Honoré Daumier,
A Surprise, 1853,
lithograph.

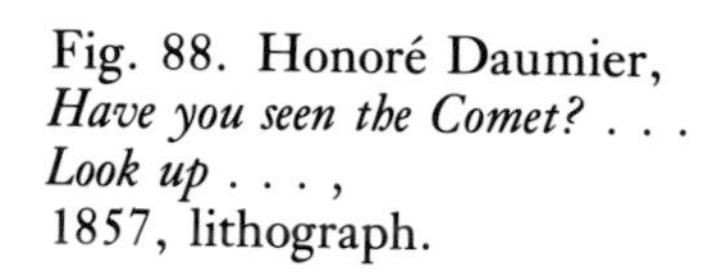

Fig. 88. Honoré Daumier,
Have you seen the Comet? . . .
Look up . . . ,
1857, lithograph.

Fig. 89. Honoré Daumier,
The German Astronomer
Releases a Famous "Canard,"
1857, lithograph.

Fig. 90. Mary Evans(?), *Donati's Comet Over the Conciergerie in Paris on October 5, 1858*, lithograph.

Fig. 91. Stephens (?), *The Great War Comet of 1861*, from *Vanity Fair*, August 3, 1863. This illustration of General Winfield Scott may have been inspired by Tebbutt's Comet, thought to have been the brightest comet of the century.

THE GREAT WAR COMET OF 1861.

The comet's appearance caused consternation among the general public and probably stimulated yet another generation of artists to employ topical comet imagery. For example, Gustave Moreau's *The Procession of the Kings* of 1861 (in the Musée Gustave Moreau, Paris) upholds the tradition of substituting a comet for the Star of Bethlehem, while an illustration for *Vanity Fair* personifies General Winfield Scott, a warmonger, as "The Great War Comet of 1861" (Fig. 91). A caricatured portrait of the general functions as the head of the comet; the general's Victorian whiskers and hair form the coma; and an army of rattling, sharply pointed bayonets comprises the comet's tail.

Tebbutt's Comet was the first comet to receive attention from a professional photographer, the Englishman Warren de la Rue, one of the great pioneers of astronomical photography. After seeing daguerreotypes at the Exhibition of 1851, he photographed the sun and eclipses. He also tried, without success, to record a comet apparition. De la Rue did manage, however, to preserve a record of Tebbutt's Comet for posterity in the drawings that he made with the aid of a telescope.

Another "Civil War comet" was the Great Comet of 1862, the periodic comet Swift-Tuttle, discovered by the American Lewis Swift. Although it failed to equal Tebbutt's Comet in either magnitude or brilliance, it was nevertheless striking and notable for the peculiar jets of light, frequently altering in their forms, that spurted from its nucleus. Some superstitious comet watchers interpreted these celestial fireworks as a herald of the battles of Shiloh and Williamsburg, while Thomas Hardy's fictional astronomer in *Two on a Tower* was saved by the very comet: "The strenuous wish to live and to

Fig. 92. *Donati's Comet Over Hamburg in October 1858*, watercolor, 14⅜ x 25¾ in. Formerly, Hamburg, F. Dörling.

Fig. 93. Grandville, *Travels of a Comet*, from *Un Autre Monde*, 1844. In this colored wood engraving, a comet is personified as a courtly woman gliding across the stage, the long train of her elegant gown decorated with stellar patterns.

behold the new phenomenon supplanting the utter weariness of existence . . . gave him a new vitality. The crisis passed. . . . The comet had in all probability saved his life. The limitless and complex wonders of the sky resumed their old power over his imagination. . . ."

On April 17, 1874, a bright comet was sighted from Marseilles by J. E. Coggia (Fig. 94). It had at least three nuclei points and at one time exhibited a star-shaped nucleus. Bright jets spurted out from its head, and a black line was seen bisecting its tail. Astronomical drawings of the time attest that its head was surrounded by many envelopes.

People never seem to tire of comets, and their appetites must be periodically assuaged by feasting on new apparitions. Toward the end of "The Comet Century," after photography was in use and the charm of Realism had somewhat waned, many artists, intoxicated by actual apparitions, once again realized the potent, emotional symbolism of comets. One case in point is Jules Verne's *Off on a Comet*, published in 1878, a year of three comets. A second is found in an illustration for another of Verne's books, *From the Sun to the Moon*.

Fig. 94. Charles La Plante, *Coggia's Comet From The Pont Neuf*, 1874, engraving.

Fig. 95. James Smetham, *Sir Bedivere Throwing Excalibur Into the Lake*, ca. 1870–75, oil, 4½ x 12 in. Formerly, New York, Sotheby's.

Many of these later examples were indebted to the fantastic images of Jean-Jacques-Isidore Gérard, called Grandville, one of the shining stars of the golden age of French illustration. For his book *Un Autre Monde*, published in 1844, Grandville conjured up his *Travels of a Comet* (Fig. 93). The setting, a majestic celestial ballroom filled with accoutrements comprised of shooting stars and comets, demonstrates the artist's inventiveness, which extended even to his monogram in the lower right-hand corner. Grandville's self-portrait appears in another illustration for the same work, where the so-called Juggler of the Universe tosses the Cross of the Legion of Honor, in the guise of a meteor or comet, to the artist holding his portfolio. Such is Grandville's fanciful comment on the fleeting nature of fame and glory. A more lugubrious note is sounded in his most famous book, *Les Animaux*, where he illustrates the moment of the turtledove's passing. The bird's noble death, due to unrequited love, is signified by a comet. Many critics view this illustration as one of Grandville's most powerful and psychological works partly because of the comet, which seems to embody the soul of the turtledove, poignantly echoing his demise.

Many comets populated the skies of the nineteenth century, in part because sophisticated telescopes made more discoveries possible, and so comet fever became even more contagious, and a virtual epidemic ensued. Furthermore, comet watching was a diverting way for the middle class to spend some of their evening hours before the invention of television.

Depictions of comets were especially popular among artists in the late nineteenth century. When Frederic Remington illustrated Henry Wadsworth Longfellow's poem "Hiawatha" in a series of paintings in 1890, he depicted Ishkoodah, the comet, as an Indian warrior with long trailing hair flying over a landscape. ("Glared like Ishkoodah, the·comet,/like the star with fiery tresses.") When James Smetham represented a scene from Tennyson's

Idylls of the King (1869) in his painting *Sir Bedivere Throwing Excalibur Into the Lake* (Fig. 95), he converted Tennyson's aurora borealis into a comet, demonstrating that he was no stranger to the literary tradition wherein comets signal the death of kings. Smetham's work illustrates the moment when Sir Bedivere, after twice failing to obey the dying Arthur's wishes to throw Excalibur into the lake, finally:

> Wheel'd and threw it. The great brand
> Made lightning in the splendour of the moor
> And flashing round and round, and whirl'd in an arch
> Shot like a streamer of the northern morn [aurora borealis] . . .
> So flash'd and fell the brand Excalibur
> But ere he dipt the surface rose an arm . . .
> And caught him by the hilt and brandished him. . . .

Another type of death and destruction is symbolized by the large comet in Gustave Moreau's watercolor *The Death of Phaeton* (Fig. 96), where the comet at the lower left holds the key to unraveling this enigmatic symbolist allegory of the human soul. (Incidentally, the comet is not present in the painting's literary source, Ovid's *Metamorphoses*.) Moreau's rich personal library, today still housed in the artist's atelier in Paris, contains many scientific and astronomical books, revealing that the artist, an inveterate bibliophile, had a substantial knowledge not only of comets in particular but of astrology and astronomy in general. In addition to more rigorous sources, he owned the important scholarly best-seller by Camille Flammarion, *Astronomie Populaire*. Equipped with his knowledge, Moreau employed star maps to lay out the composition of his *Phaeton*, in which a comet traverses the constellations Leo, Hydra, Eridanus, and Auriga (the charioteer who loses his chariot, like Phaeton). This comet parallels Phaeton's own incendiary demise after he defies his father, Apollo, and makes his futile attempt to drive the chariot of the sun. The blurred contours of this comet, which seem to replace Zeus's thunderbolt in Ovid's narrative, suggest that Moreau had observed actual comets—perhaps Donati's in 1858, Tebbutt's in 1861 or subsequent apparitions. Two comets, including the second apparition of the faint Comet Encke, were in fact sighted by naked eye in 1878, the very year that Moreau painted *Phaeton*.

In the 1880s, several remarkable comets appeared. Among them was the Great Southern Comet of 1880, with its forty-degree tail extending from below the horizon. On May 22, 1881, Tebbutt spotted another comet from Windsor, Australia, which was photographed by several people. But it is Sir David Gill, the British astronomer who observed at the Cape of Good Hope, who is credited with first successfully photographing a comet—the Great

Fig. 96. Gustave Moreau, *The Death of Phaeton*, 1878, watercolor, 39 x 25½ in. Paris, Musée du Louvre, Cabinet des Dessins.

Comet of September 1882 (Fig. 67). It was initially spotted in the southern sky on September 1 by sailors aboard an Italian ship. Because of its long, sweeping tail, identified since ancient times as the Turkish Scimitar, the Great Comet of 1882 was of the type that traditionally struck terror into people's hearts. Adding to the drama, its tail had multiple disruptions, and a portion of its mass separated from the parent body to form an independent satellite, a feature that made it especially compelling to the masses. The Great Comet of 1882 was also a daylight comet, visible to the naked eye for five and a half months and remaining in telescopic range for nine, encouraging both scientific study and Romantic ruminations.

The importance of the new photographic technology dawned on cometologists, who realized that they could thus preserve evidence of apparitions which otherwise might have been lost or undiscovered. Later in 1882, a photograph of a total eclipse visible from Egypt revealed not only the corona of the sun but also a bright comet close to the sun. This mysterious stranger had never been previously seen, nor was it ever spotted again. These photographs are therefore irreplaceable as the only record of Tewfik's Comet (named after the ruler of Egypt at that time).

By the 1890s, scientific photographs of comets became quite common. While it was the English and American amateurs who had initially led the field in astronomical photography, the professionals soon recognized its potential when coupled with the use of a telescope. Long exposures made it possible to observe and chart stars along with other bodies that were either too far or too faint to spy through a telescope with the eye alone. The camera thus opened up a whole new universe of possibilities for cometologists. It also equipped scientists for an unparalleled comprehensive study of the 1910 apparition of Halley's Comet.

In no other century of Western history did comets carry such diversified meanings as they did in the nineteenth. Since science had removed the stigma and most of superstition's sting surrounding comets, they could be regarded humorously, even capriciously, as we have seen. Scores of publications called *The Comet* appeared, and then, true to their namesake, disappeared as quickly. Even *Puck*, the witty periodical dealing primarily with social issues, was published in the United States under the impish aegis of a comet. In 1841, *The Comet of Many Tails: a comic kalendar* was issued. The public's love affair with comets led some individuals to proudly name ships *Metacomet* or *The Blazing Star*, but it also fed scandalmongers in the infamous case of the *Comet*, a steamboat whose end just had to be catastrophic. The headlines regarding its fate read: ACCOUNT OF A MOST MELANCHOLY AND DREADFUL INCIDENT! LOSS OF THE COMET STEAMBOAT. 70 PERSONS DROWNED.

During the nineteenth century, comets still frequently carried symbolic import and continued to appear in their old guises as on coats of arms, for instance, as preserved in a detail from a magnificent Vatican pavement (Fig. 97). They also figured in the historicism that pervaded the century, as in

Fig. 97. Coat-of-arms of Pope Leo XIII (1878–1903), inlaid stone from a floor pavement. Rome, The Vatican.

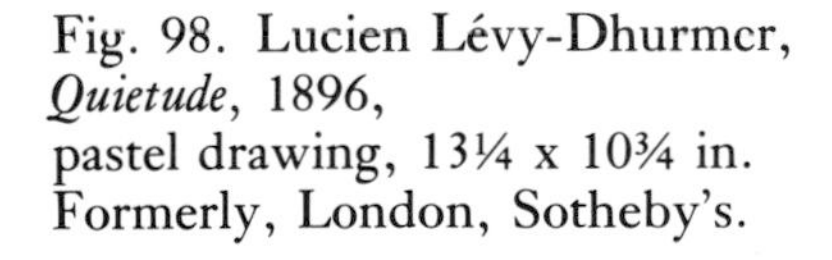

Fig. 98. Lucien Lévy-Dhurmer, *Quietude*, 1896, pastel drawing, 13¼ x 10¾ in. Formerly, London, Sotheby's.

Lucien Lévy-Dhurmer's pastel *Quietude* (Fig. 98), a reworking of the theme of Dürer's *Melencolia I* (Fig. 31) drawn in the background as an historical footnote.

Yet despite all the technical advances, comets were never entirely stripped of their mythology during this comet-crazy century. It ended with neither a spectacular comet (though five comets were sighted by telescope in 1899) nor a bang, but with a click—the noise of a camera lens, signifying the potential that medium held for astronomy, space exploration, and of course, for discovering the true nature of comets, a quest which even today remains incomplete.

> Quit his Comet! Never! His Comet was his castle. . . .
>
> (Jules Verne, *Off on a Comet!*)

Six

The Twentieth-Century Escape

A comet could be the perfect emblem for the twentieth century, for comets have traditionally signaled change. At the turn of the century, there was a pervasive feeling that something had gone wrong with European civilization. Intimations of profound change and political revolution were in the air. Some people consequently buried themselves in hedonism, but no one could deny that Nietzsche had proclaimed that God was dead and Malthus's doctrine of overpopulation was fast becoming a reality. Unorthodox nineteenth-century artists, philosophers, and writers had attacked the Victorian taste for outward convention, conformity, and order. Emancipation of workers and women had begun, in tandem with a general democratization of society.

In progressive artistic circles, such changes produced an exhilaration whose momentum was reflected in H. G. Wells's utopian, futuristic book *In The Days of the Comet*. A clear writing style had cut through the verbosity of the previous age. Art for art's sake, a nineteenth-century aesthetic trend, with its patron saints Charles Baudelaire and Walter Pater, had paved the way for abstract art. The accelerated rate of scientific discoveries and their implications—those of Einstein's new physics, for instance—were heady but frightening. This new influx of information argued that beneath visual reality as we know it lurk hidden, subtle forces never before imagined. Traditional styles and images that had corresponded to a stable, orderly universe, in which thought and expression were tamed and disciplined, were thrown away as invalid.

Since major art patrons were no longer the Church and state, artists could retreat into their own private worlds, creating images for themselves alone. Henri Bergson's writings reflect these subjective developments, and the loss of absolute standards. In *Creative Evolution*, published in 1907, he prophetically referred to the relativism that recent scientific discoveries had sug-

Fig. 99. Wassily Kandinsky, *Comet (Night Rider)*, 1900, watercolor and gold-bronze on red paper, 7¾ x 9 in. Munich, Städische Galerie im Lenbachhaus.

gested; he also helped to rehabilitate the concept of intuition, pointing out the forces of instinct and chance, and the superiority of metaphysical over merely physical truth. This attitude encouraged an escape from reality toward imaginative new worlds—fertile territories for comet images.

Wassily Kandinsky's painting *Comet* (*Night Rider*) of 1900 (Fig. 99) seems to straddle the two centuries. Its decorative, sinuous forms and strong palette grow out of the German Jugendstil movement, the Art Nouveau of Munich, and foreshadow the simplified abstract forms of the artist's later style. While the motifs of the castle and knight on horseback were derived from the nineteenth-century Romantic vocabulary, Kandinsky adopted the rider as both his own personal symbol of liberation and as the logo for the pivotal German Expressionist group to which he belonged, *Der Blaue Reiter* (The Blue Rider). Kandinsky's huge comet sweeping across the sky suggests the mystery of cosmic forces, which always intrigued the artist, as noted in his famous essay "On the Spiritual in Art." His comet also seems to presage change, for the world and for art. Kandinsky created many forms that may well have been inspired by comets, as in his *Small Worlds* series, filled with nonrepresentational shapes of this nature.

One must be cautious in assigning representational identities to the forms used in nonrepresentational works of art. Because abstract works tend to be suggestive, and thus create definitive associations for the human mind, there is a problem in analysis that is pervasive in dealing with the art of the twentieth century. Gustav Klimt's *Beethoven Frieze* is a case in point. A detail (Fig. 100) shows two embracing lovers over whom hover two cometlike or vegetal forms, one male and the other female. They are evocative of joyous sexual energy and reproductive potential. Since comets and shooting stars have been allied with love and sexual union as far back as Elizabethan times, these

Fig. 100. Gustav Klimt, detail of the *Beethoven Frieze*, 1902, casein paint on stucco, inlaid with semiprecious stones, 86⅝ in. high. Geneva, Erich Lederer Collection (stored in Vienna, The Lower Belvedere).

Fig. 101. Arthur Rackham, *To Hear the Sea-Maid's Music*, illustration to *A Midsummer Night's Dream*, 1908, watercolor, 15 x 10½ in. London, The Victoria and Albert Museum.

Arthur Rackham 1908
NON
SINE SOLE
IRIS

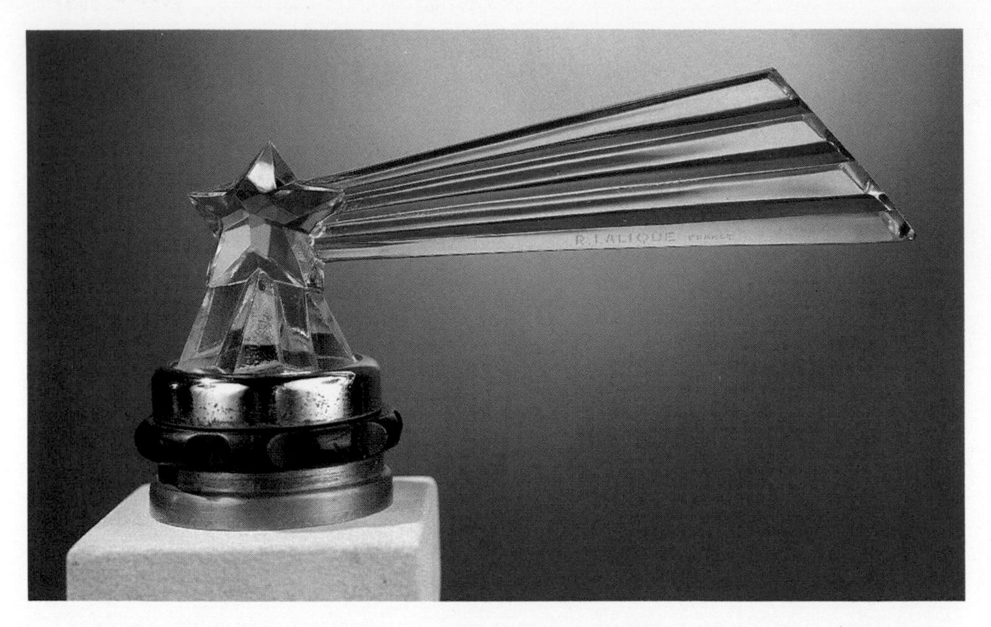

Fig. 102. René Lalique, *Comète,* car hood ornament, 1927, glass, tail measures ca. 6 in. André Surmain Collection.

shapes may be derived from comets. As Edmund Spenser wrote in "Shepherd's Calendar":

> Tho gan my lovely Spring bid me farewel,
> and summer season sped him to display
> (For love then in the Lyons house did dwell)
> The raging fyre, that kindled at his ray.
> A comett stird up that unkindly heate,
> That reigned (as men sayd) in Venus seate.

Arthur Rackham echoes this tradition in his watercolor illustrations to *A Midsummer Night's Dream*, published by William Heinemann in 1908. The illustration to Act II, Scene 1, of Shakespeare's play features a poetic celestial spectacle not unlike a meteor shower (Fig. 101). This fall of stars is directly related to armed Cupid, drawn by the artist in a roundel at the top of the composition. The god of love can, therefore, be interpreted as releasing the meteors like arrows.

During the twentieth century, there have been several hundred comets sighted, but very few by the naked eye. The first memorable one, a brilliant daylight comet, appeared in 1910. It was followed in the same year by the long-awaited return of Comet Halley, which probably inspired the following lines by James Joyce, from *Finnegans Wake*: "Any dog's life you list you may still hear them at it. . . . as ever sure as Halley's comet."

While astronomers continued to search the heavens with ever more sophisticated instruments and understanding, there was a lengthy hiatus in naked-eye apparitions. Not until 1927, with Comet Skjellerup, which was bright for only a few nights, were their expectations rewarded. Thereafter, another hiatus occurred until the 1940s, when several comets approaching the sun were visible without use of a telescope.

During this period of few spectacular apparitions, however, artists continued to create comet images. Due to the disillusionment and apocalyptic climate produced by World War I, comets once again functioned as portents or, alternatively, as vehicles of escape, often associated with nature. The former trend is reflected very graphically in the violently expressionistic painting by Ludwig Meidner, in which a turbulent comet in the upper left shines glaringly down upon a scene of horror and devastation. Meidner's forms and mood mirror the medieval and Renaissance German tradition of Apocalypse illustrations (Figs. 5, 32 and 33). A painting by the American artist Charles Burchfield, *Star and Fires*, executed around 1920, also has a very apocalyptic feeling, while his earlier watercolor, *House and Tree by Arc Light* (Fig. 104) suggests not only the mystery of cosmic forces but also the sense of wonder they create. Other examples occur in the works of the German artist Franz Marc, whose abstract oil painting of 1914, *Fighting Forms*, has recently been linked in its mood to medieval and Renaissance illustrations of Revelation. The German expressionist image of destruction and carnage was foreshadowed in a poem of 1912 by Georg Heym entitled "Umbra Vitae":

> The people on the streets draw up and stare
> While overhead huge portents cross the sky;
> Round fanglike towers threatening comets flare,
> Death-bearing, fiery snouted where they fly. . . .
>
> Through night great hordes of suicides are hurled.
> Men seeking on their way the selves they've lost.

In 1917, Marc's drawings for *Stella Peregrina*, based on earlier watercolor illustrations of 1906 for a poem by Gustav Renner, were published posthumously (Fig. 103). He had been killed in battle in 1916. The wandering star of the title expresses his youthful fascination with nature and cosmic themes, while the plates themselves harken back to the German woodcut tradition, a frequent source for the German expressionists. Marc sought to suggest the cosmic unity of all things at a deeper, more spiritual level and, like Kandinsky, belonged to *Der Blaue Reiter*. He wrote that he wanted "to get back . . . to the mysterious and abstract images of inner life." His language is reminiscent of the concurrent interest in psychoanalysis, the collective unconscious, and symbols in general. He also employed a comet for an *Ex Libris* he designed.

As in previous ages, comet imagery generally appealed to artists with a penchant for bizarre or esoteric symbolism. An excellent example is Fidus's *Gottsucker* (Fig. 105), whose eclectic yet novel image depicts a male nude plummeting through star-studded space with red rays of fire behind him. His

Fig. 103. Franz Marc,
Dort Fiel Ein Stern,
from *Stella Peregrina*, 1917.

Fig. 104. Charles Burchfield,
House and Tree by Arc Light,
1916, watercolor,
20 x 13⅞ in.
Utica, New York,
Munson-Williams-Proctor
Institute.

Fig. 105. Fidus (Hugo Höppner),
Der Gottsucker,
1918, oil on paper, 25¾ x 35 in.
Formerly, London, Sotheby's.

Fig. 106. Maurice Guiraud-Rivière,
The Comet, ca. 1930,
gilded and painted bronze,
21⅞ in. high including base.
Formerly, New York, Christie's.

body and its flaming trajectory are echoed by the plunging heart-shaped forms with their fiery tails on the lateral borders of the work. Fidus's image recalls the myth of Icarus, who flew too near the sun. It thus suggests that, in seeking God, the figure reached too high only to fall to Earth like a comet, or more probably a shooting star. Because Fidus depicted the figure reaching upward, the painting would seem to be an image of hope, rather than of empty despair. In these characteristics, it is typical of decadent *fin de siècle* escapist themes, which blended idealism and mythology with images of despair.

Art deco, that svelte style of the late 1920s and 1930s whose celebration of modern glamor was based on geometry and chic, was also a form of escapism. Since the style embraced heavenly images, such as stars and the moon, it is no wonder that it also spawned a number of streamlined comets. One of the most beautiful is René Lalique's glass car-hood ornament (Fig. 102), a charming marriage of form and function. A comet once again takes the form of a woman with flowing locks in Maurice Guiraud-Rivière's bronze sculpture (Fig. 106). The French designer Paul Iribe—an associate of Paul Poiret—who created furniture and the famous silhouette logo for Lanvin, designed several pieces of jewelry using comet forms, among them a graceful, dazzling diamond necklace for Coco Chanel, the French fashion doyenne. Its

Fig. 107. Paul Iribe, Comet necklace for Coco Chanel, ca. 1932. Paris, Bibliothèque Forney.

Fig. 108. Joan Miró, *Landscape (The Hare) Autumn*, 1927, oil, 51 x 76⅝ in. New York, The Solomon R. Guggenheim Museum.

tail sinuously drapes around the back of the wearer's neck to cascade down the chest (Fig. 107).

The disillusionment experienced in the wake of World War I expressed itself most vociferously in the artistic-literary movement called Dada. In the 1920s, the leaders of Dadaism rode the phoenix of Surrealism to the subterranean currents of the human mind. They depended on such aids as automatic writing, dreams, hallucinations, and drugs to evoke new worlds and to tap unusual symbols in the human unconscious. André Masson, an artist who was traumatized by his experiences in the trenches during World War I, produced many works with images suggestive of comets and other whirling cosmic forms, as in his oil painting *Landscape of Wonders* of 1935. His contour drawing *Mithra* of 1933, published in 1936 in a portfolio series of etchings (Fig. 109) entitled *Sacrifices, The Gods Who Die,* fits into Masson's vision of nature as a kind of cosmic madness. The series includes depictions of the sun, stars, the Minotaur, and the labyrinth, a surrealist symbol of weariness. Masson's *Mithra* alludes to the Roman mystical god who slayed the cosmic bull, thereby releasing the beneficent forces of the earth. Masson placed a swirling comet in the upper left-hand section of his composition, indicating

Fig. 109. André Masson,
Mithra, from *Sacrifices*,
1936, etching.

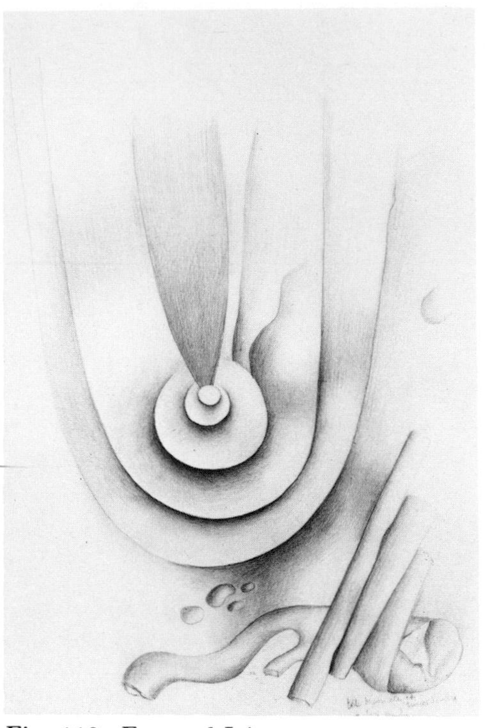

Fig. 110. Fernand Léger,
Head of a Comet and Trunk of a Tree, 1931,
black chalk drawing, 17⅛ x 11⅝ in.
Formerly, New York, Sotheby's.

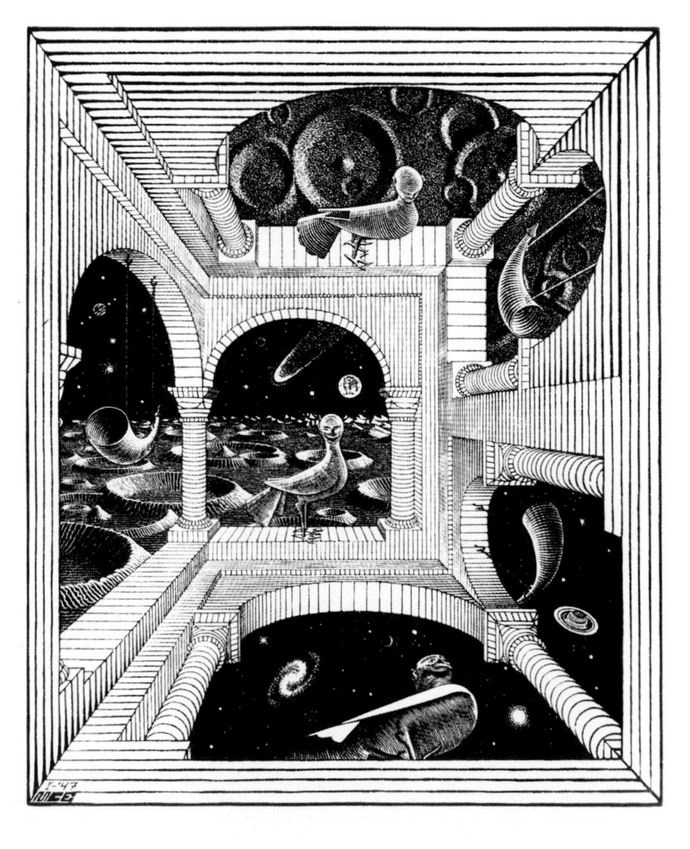

Fig. 111. Maurits Cornelius Escher,
Other World, 1947,
wood engraving and woodcut
printed in three colors.

that he was familiar with James George Frazer's eclectic anthropological book on dying fertility gods, *The Golden Bough*, as well as the ancient Roman visual tradition wherein Mithra is represented in the company of comets (Fig. 24). In the same year, Masson also designed a backdrop for the ballet *Les Présages* at the Monte Carlo Theater in Paris, with its choreography by Léonide Massine and a score from Tchaikovsky's Fifth Symphony. Masson's backdrop featured all sorts of dramatic heavenly turbulence—whirling suns, shooting stars, comets—as symbols of cosmic instability in a sky marked by brilliant zones of color.

Giorgio de Chirico, who was also influenced by Frazer's *The Golden Bough*, designed costumes for Diaghilev's surrealistic ballet *Le Bal* in 1929; his costume design for the astrologer features two comets on the coat tails. Joan Miró, another surrealist artist, also employed suggestive cometlike forms in his works (Fig. 108), but they are usually more generalized and personal than those of Masson. The inscription of Fernand Léger's drawing of a comet identifies the subject matter precisely (Fig. 110). Interestingly, Léger, who tempered his early cubism with surrealist overtones, has rendered concentric envelopes of light around the comet's head in this image, indicating that he probably consulted scientific drawings of comet envelopes or actually studied a comet firsthand. The verso of the drawing shows abstract geometrical forms derived from the more descriptive representation of the comet.

The Surrealists were often concerned with sexual themes and symbols. Dorothea Tanning's painting *The Truth About Comets and Little Girls* of 1945 (Fig. 113) suggests in a typically ambiguous fashion the expectancy of adolescence and coming of age. This transitional scene is set in a frigid landscape with a stairway to nowhere. In this context, two large comets seem to function as sexually explosive symbols that arabesque across the wintry sky. The artist has long been fascinated with these celestial forms, and related that she and Max Ernst, another surrealist artist, would spend nights searching the Arizona desert skies for meteors and comets.

A more accessible—and nostalgic—representation of a comet occurs in Frank Tenney Johnson's *An Evil Omen* (Fig. 114), wherein the Native American riders regard a comet streaking across a vivid turquoise sky. The men embody the theme of the noble savage in harmony with nature, while the comet presages the downfall of their once-proud civilization.

World War II followed all too quickly on the heels of the earlier war, intensifying the escapist use of comet imagery in art. A comet streaks across the sky in one of M. C. Escher's enigmatic, highly manipulated perspectives where gravity seems not to exist (Fig. 111). The comet adds a disquieting note, but also a sense of strange familiarity in a visionary universe that appears alien yet tantalizing; to quote J. B. S. Haldane, the British biologist, "My own suspicion is that the universe is not only queerer than we suppose, but queerer than we can suppose." Escher's following comment on his own works is most revealing:

> I am so far removed from the beautiful earth that the people scurrying around in their cities are of course no longer visible. Whether this is such a good thing as it seems to be at the time is open to question. But it is a fact that my vision is so beautiful, so restful and so full of peace that I usually fall asleep soon afterwards.

The existential postwar world was a place where human beings were condemned to terrifying loneliness and despair; only action validated existence. This atmosphere exacerbated the need for artists to create their own worlds and expression. Rufino Tamayo, a Mexican artist, was one of these people interested in visionary worlds and themes. In *The Astrologers of Life* (Fig. 112), two silhouetted figures search the skies with telescopes, while a comet pierces the azure blue expanse behind their backs. Blueprints with scientific diagrams and geometric forms are strewn across the foreground. In the background, a red radio tower perches on a barren rocky promontory and sends out signals, perhaps meant to express people's inability to read the signs and their futile attempt to understand the workings of the universe with their technology. Tamayo's paintings abound with comets, among them *Cataclysm*, *Stargazer*, and *Man Confronting Infinity*, a large mural done in 1971. Tamayo, who was also knowledgeable in Pre-Columbian astronomy, was one of six artists invited to NASA headquarters in Washington, D.C., to meet with scientists to discuss the theoretical relationship between art and science.

The sculptor Theodore Rozsak had a lifelong interest in cosmic themes of creation and destruction. His drawing *Study—Meteor* of 1962 (Fig. 115), which actually looks more like a comet, was never translated into metal but reflects the escapist mood of the early 1960s as well as the artist's fascination with cosmic energy. A similar harsh mood is created in the following lines of M. L. Rosenthal's poem "If I Forget Thee . . . ," which effectively uses comet imagery:

> Honor the poised sword of man's misery; the faces
> of the mothers; the cobblestones
> on which men walk to war and no fortune.
>
> Honor not that light, cold light from an unknown star,
> light that affirms nor denies, falling across the fields,
> whiting the stones, whetting the wind.
>
> Curve of the comet, or peal of distant sound,
> or friend who can no longer turn to me:
> take me not, carry me not, show me no light of star. . . .

Fig. 112. Rufino Tamayo, *The Astrologers of Life*, 1947, oil, 80 × 60 in. Formerly, New York, Sotheby's.

The experiments with rockets during World War II led to the serious exploration of space and the decades of Sputnik and other satellites. In 1957, two nonperiodic comets appeared—Comet Arend-Roland, with its frontward spike (an unusual phenomenon, which Whipple suggests may result from a newly formed, unbaked comet), and Comet Mrkós. During the succeeding years, exploration of space continued, but there were disappointingly few memorable comets, save those of the sungrazer Comet Ikeya-Seki in 1965, Comet Bennett in 1970 (Fig. 1), and Comet Kohoutek in 1973–74 (which, except for the Skylab experiments, would be easily forgettable). Kohoutek was actually observed by U.S. astronauts outside their craft during a space walk 270 miles above Earth. The last spectacular comet since then was Comet West in 1976 whose breakup was photographed. Its head divided into

Fig. 113. Dorothea Tanning, *The Truth About Comets and Little Girls,* 1945, oil, 23½ x 23½ in. Chichester, Edward James Foundation.

Fig. 114. Frank Tenney Johnson, *An Evil Omen*, 1930, oil, 24 x 18 in. Tuscaloosa, The Warner Collection of Gulf States Paper Corporation.

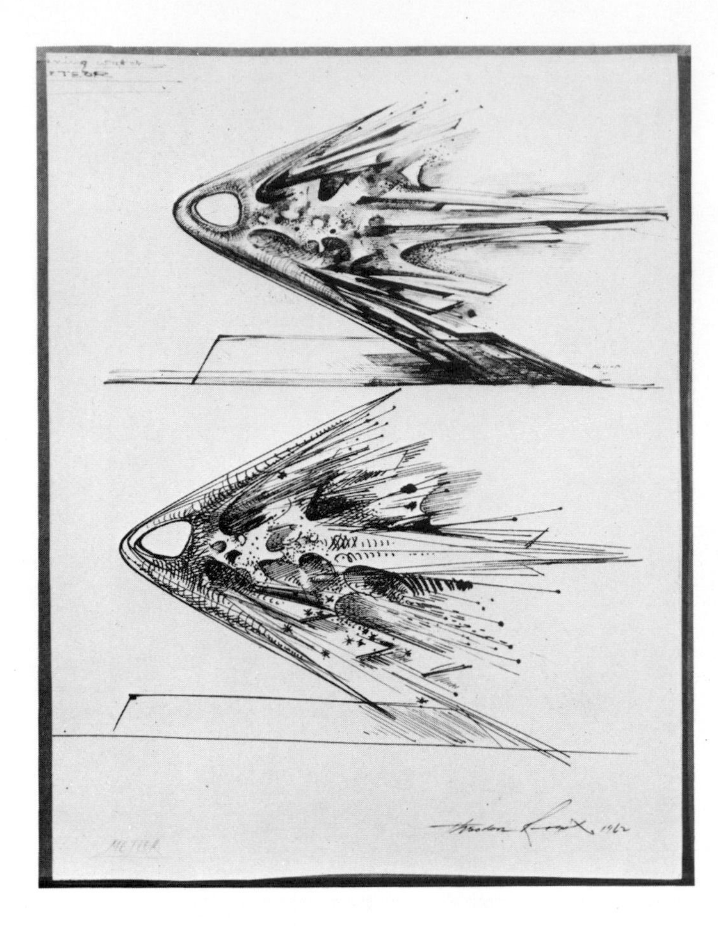

Fig. 115. Theodore Roszak, *Study—Meteor*, 1962, ink drawing, 11¼ x 8½ in. New York, Collection of Whitney Museum of American Art, gift of Sara-Jane Roszak.

four parts and then threw out streamers of dust every two or three days, creating a fanning tail.

The twentieth century, which has worshipped novelty and the speed of change, has also embraced ambiguity. In the visual arts, as forms are abstracted, they become less identifiable. This effect creates problems for comet seekers. Any form with a tail could potentially be interpreted as a comet or shooting star. Take, for example, the painting *Actual Size* by the American Edward Ruscha (Fig. 116). While his forms do suggest the curving tail and trajectory of a comet, the letters SPAM clearly state that the image is not a comet. While the painting cleverly reeks of commercialism, advertising, and the banality of much of pop-art subject matter, the pure form does, however, suggest the excitement, movement, and transitory nature of a comet.

In the heterogenous art world of the 1980s, artists continue to create comet images. The most realistic are those by artists who specialize in illustrating aspects of space, such as the German Rudolf Brammer (Fig. 22), and the painters of historical exactitude, such as William Davis, the contemporary American artist who painted the nineteenth-century steamboat *Metacomet* (Fig. 117). Then there are the symbolic artists growing out of surrealism who incongruously juxtapose objects like the massive egg and comet in Alan Magee's monotype of 1984 entitled *Comet*. At the other end of the spectrum are the visionaries and the artists, such as Jim Nutt, whose work is so personal and convoluted in its symbolism that any cometlike form may never be definitely identified. Several artists have stated that they plan to celebrate the return of Comet Halley in 1985–86 with special works which they are creating for the event.

Only during the twentieth century have people gained enough emotional distance to really become sentimental about comets. There are scores of individuals who witnessed Comet Halley in 1910 and eagerly await its return in 1985–86 ("two-timers"). Some even state quite blatantly that it is the goal of their lives. A group of these confessions and reminiscences have been published in *Halley's Comet Watch Newsletter*. Other individuals lovingly remember the awesome event as though it happened yesterday. Katherine Burton, a retired English professor, has written:

> I remember this very well. I was four years old, held up to a window and told to remember what I saw because I would be an old lady when I had another chance to see it. It was amazing to me to think of myself as an old lady, and I swear I simply photographed the scene in my mind. I was never in that house after I was six but I could go to it and point out—from the very window—the very spot in the sky where I saw the comet. You might be interested in all that because your daughter is coming right along to the age where she'll be picking up lifelong memories.

Will the chain continue? In 2061, will my daughter Allegra, at the age of seventy-nine, gaze at Comet Halley and remember?

Fig. 116. Edward Ruscha (b. 1937)
Actual Size,
1962, oil, 72 x 67 in. Los Angeles, Los Angeles County Museum of Art, anonymous gift through The Contemporary Art Council.

Cometomania is alive and well. The twentieth century has produced an unfathomable amount of comet memorabilia; comets decorate postcards, playing cards, stained-glass windows, shoes, nightgowns, video games, and advertising of all sorts. Cometologists can subscribe to a serious periodical with the most up-to-date information on comets, *The Comet News Service*, edited by Dr. Joseph N. Marcus. There are also a growing number of comet clubs, including Halley's Comet Society, founded in 1976 in London. Several exotic cruises are planned to the Amazon River and other spots in the Southern Hemisphere for the most advantageous viewing of Comet Halley in 1986. Even as I write, medals, T-shirts, buttons, mugs, and other items emblazoned with Halley's famous fiery profile, along with various slogans, are being manufactured and marketed to commemorate its coming apparition.

It is easy to understand why comets are symbolically charged images and why the very word *comet* connotes magic. Even the group Bill Haley and the Comets rock-and-rolled to notoriety with a little help from their catchy (if mispronounced) name. The wondrous word has even been attached to such mundane objects as cleansers, chocolates, ice-cream cones, and automobiles, lending them a miraculous, seductive attraction. There is something inherently exciting about Meteor Beer or the Blazing Star Ferry or the Comet

Fig. 117. William Davis, *The Steamboat Metacomet*, 1984, acrylic, 20 x 32 in. Dennis, Massachusetts, Hyland Granby Antiques Collection.

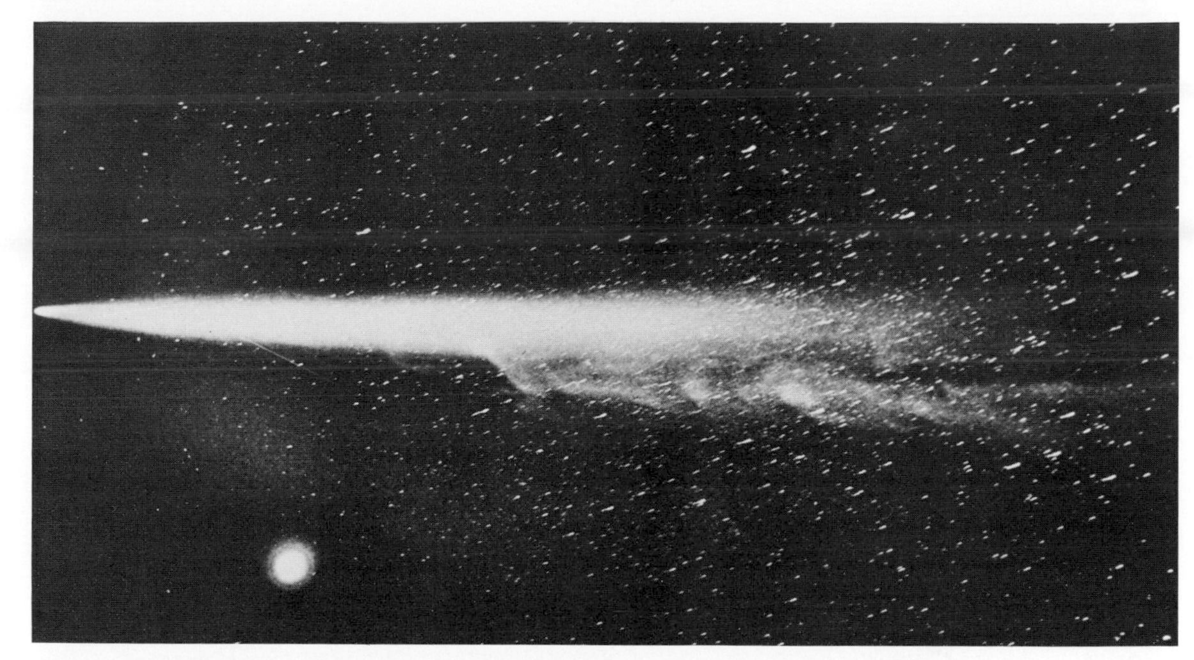

Fig. 118. *Comet Halley*, 1910. The bright light below the comet is the overexposed image of the planet Venus. Lowell Observatory photograph.

Travel Agency and something undeniably star-crossed and magnetic about the rue de la Comète in Paris or the Teatro della Cometa in Rome. Comets are mythical.

In spite of recent triumphs in science, it seems that human nature has not changed much in two thousand years, but other factors in our environment and society have altered. Human beings now stand on the razor's edge, possessing an unforetold potential to destroy themselves completely—not only spiritually, with apathy, cynicism, and disillusionment, but quite literally as well, with deadly weapons. Comets, by their very existence, argue against this bleak scenario. They can be understood not as portents of good or evil but as eloquent symbols of change within the continuum of our universe. They quietly argue for the preservation and improvement of our planet, while offering scientists elemental clues to the formative process of the universe. They are beacons of hope and inspirations for the discovery of new worlds.

Selected Bibliography

Brandt, John C., ed. *Comets: Readings from Scientific American*. San Francisco, W. H. Freeman and Co., 1981 (republications of significant articles, including a number by Whipple).

Brandt, John C., and Chapman, Robert D. *Introduction to Comets*. Cambridge, Cambridge University Press, 1981 (contains an excellent, extensive bibliography).

Brown, Peter Lancaster. *Comets, Meteorites and Men*. New York, Taplinger, 1974.

Calder, Nigel. *The Comet Is Coming!* New York, The Viking Press, 1980.

Datei, Enea. "La Cometa di Halley in un affresco del XIII sec.?" *Astronomia*, 23, June 1983, pp. 30–34.

Hellman, C. Doris. *The Comet of 1577: Its Place in the History of Astronomy*. London, P. S. King & Staples, Ltd., 1944.

Hess, W. "Himmels und Naturscheinungen in Einblattdrucken des XV bis XVIII Jahrhunderts." *Zeitschrift für Bucherfreunde*, 2, Leipzig, 1911, pp. 353–404.

Hollaender, Eugen. "Wundergeburt und Wundergestalt in Einblattdrucken des fünfzehnten bis achtzehnten Jahrhunderts." *Bieträge aus dem Genzgebiet zwischen Medizingeschichte und kunst-kulturen-Literatur*, 4, Stuttgart, 1921.

Lyttleton, R. A. *The Comets and Their Origin*. Cambridge, Cambridge University Press, 1953.

Massing, Jean-Michel. "A Sixteenth-Century Illustrated Treatise on Comets." *The Journal of the Warburg and Courtauld Institutes*, 40, 1977, pp. 318–22.

Moore, Patrick and Mason, John. *The Return of Halley's Comet*. Cambridge, Patrick Stephens, 1984.

Olson, Roberta J. M. ". . . And They Saw Stars: Renaissance Representations of Comets and Pretelescopic Astronomy." *The Art Journal*, 44, Fall 1984.

———. "Giotto's Portrait of Halley's Comet." *Scientific American*, 240, May 1979, pp. 160–70.

———. "Quand passant les cometes. . . ." *Connaissance des Arts*, 380, October 1983, pp. 72–77.

Richter, N. B. *The Nature of Comets*. New York, Dover, 1963.

Ronan, Colin A. *Edmond Halley: Genius in Eclipse*. New York, Doubleday, 1969.

Saxl, Fritz. *Verzeichnis astrologischer und mythologischer illustrieter Handschriften des lateinischen Mittelalters*, 3 vols. in 4. Heidelberg, C. Winter, 1915–53.

Thorndike, Lynn. *Latin Treatises on Comets Between 1238 and 1368 A.D.* Chicago, University of Chicago Press, 1950.

Vsekhsvyatskii, S. K. *Physical Characteristics of Comets*. Jerusalem, Israel Program for Scientific Translations, 1964.

Warburg, Aby. "Heidnisch-antike Weissagung in Wort und Bild zu Luthers Zeiten." *Gesammelte Schriften*, 2, Leipzig, 1932, pp. 490–558, especially.

Whipple, Fred L. "The Nature of Comets." *Scientific American*, 230, February 1974, pp. 48–57.

Wissowa, Georg, ed. *Paulys Real-Encyclopädie der Classischen Altertumswissenschaft*, 11. Stuttgart, J. B. Metzlerschersche, 1922, pp. 1143–93.

Zinner, Ernst. *Verzeichnis der astronomischen handschriften des deutschen kulturgebietes*. Munich, C. H. Beck, 1925.

Illustration Credits

The author wishes to thank the following persons and institutions for permission to reproduce the illustrations below. Due to space limitations, the collections of prints, books, and photographs are only cited in the Illustration Credits and not in the captions, and media descriptions in the captions have been abbreviated.

Collections of Her Majesty, the Queen, copyright reserved: Frontispiece, Fig. 79

The University of London, The Library of the Warburg Institute: p. vii

Claude Nicollier: Fig. 1

From "Giotto's Portrait of Halley's Comet," (May 1979) drawn by Dan Todd, copyright © 1979 by *Scientific American, Inc.* All rights reserved: Fig. 2

David Wool, illustrator: Fig. 3

Courtesy of the Trustees of the British Museum, London: Figs. 4, 64, 65, 66, 73, 74, 75, 84

The Metropolitan Museum of Art, copyright ©, all rights reserved: Fig. 5 (Fletcher Fund, 1919), Fig. 31 (Dick Fund, 1943), Figs. 70, 71 (Gift of Miss Georgiana W. Sargent in memory of John Osborne Sargent, 1924)

The Research Libraries, New York Public Library, Astor, Lenox, and Tilden Foundations:
Print Collection Figs. 6, 46, 49, 54, 63, 91, 93 (Photo, Philip Pocock)
Rare Books and ManuscriptsDivision: Figs. 13, 30, 48, 60, 76;
Spencer Collection: Fig. 38;
Science and Technology Research Center: Figs. 51, 53

Library Services Department, American Museum of Natural History, New York: Figs. 7, 18, 69, 81, 90, 94

Mount Wilson and Las Campanas Observatories, Carnegie Institution of Washington, Pasadena, California: Figs. 8, 21

National Portrait Gallery, London: Fig. 9

Staats und Stadtbibliothek, Augsburg: Fig. 10

Courtesy of Birmingham Museums and Art Gallery: Fig. 11

Ville de Bayeux: Fig. 12

The Mastèr and Fellows of Trinity College, Cambridge: Fig. 14

SCALA/Art Resources, New York: Figs. 15, 16, 44

Bürgerbibliothek in the Zentralbibliothek, Lucerne: Figs. 17, 29

National Air and Space Museum, Smithsonian Institution: Fig. 19

Bibliothèque de L'Arsenal, Paris (Photo, R. Lalance): Fig. 20

Rudolf Brammer: Fig. 22

Bayerisches Nationalmuseum, Munich: Figs. 23, 59

Civici Musei e Gallerie di Storia e Arte, Museo Archeologico e Gabinetto Numismatico, Udine: Fig. 24

F.lli Manzotti: Fig. 25

Giraudon/Art Resources, New York: Figs. 26, 72

By permission of The British Library, London: Fig. 27

Photos, Vatican Museums, Rome: Figs. 28, 97

Kupferstichkabinett, Berlin SMPK: Figs. 32, 33

University of London, The Warburg Institute: Figs. 34–37

Print and Drawing Department, Zentralbibliothek, Zurich: Figs. 39, 40

The Central Library, Istanbul University: Fig. 41

Pictorial and Archive Collection of the Science Museum Library, South Kensington, London: Fig. 42

Biblioteca nacional, Madrid: Fig. 43

Documentation Photographique de la Réunion des Musées Nationaux, Paris: Figs. 45, 96
Rare Book and Manuscript Library, Columbia University, New York: Figs. 47, 50, 55, 61
The Treasury and The National Trust at Petworth House, Surrey, England: Fig. 52
Atlas van Stolk, Rotterdam: Fig. 56
Historisch Museum, Rotterdam: Fig. 57
Courtesy of Staatliche Graphische Sammlung, Munich: Fig. 58
Courtesy of Christie's, New York: Figs. 62, 106
Courtesy of Patrick Moore: Fig. 67
Photos, Bibliothèque Nationale, Paris: Figs. 68, 82, 87, 88, 89
The Tate Gallery, London: Figs. 77, 86
National Museum of Wales, Cardiff: Fig. 78
Staatliche Kunstsammlungen, Dresden: Fig. 80
Collection, The Museum of Modern Art, New York: Fig. 83
Yale Center for British Art, Paul Mellon Collection, New Haven: Fig. 85
F. Dörling, Hamburg: Fig. 92
Sotheby Parke Bernet Inc., New York copyright © 1985: Figs. 95, 98, 105, 110, 111, 112
Städtische Galerie im Lenbachhaus, Munich: Fig. 99
Österreichische Galerie, Vienna: Fig. 100
Courtesy of the Trustees of the Victoria and Albert Museum, London: Fig. 101
André Surmain, Mougins, France: Fig. 102
Princeton University Library: Fig. 103
Munson-Williams-Proctor Institute, Utica, New York. Edward W. Root Bequest: Fig. 104
Document Bibliothèque Forney, Paris: Fig. 107
The Solomon R. Guggenheim Museum, New York, photo by Carmelo Guadagno and David Heald: Fig. 108
Alderman Library, The University of Virginia, Charlottesville: Fig. 109
Executors of the Estate of Edward F.W. James, deceased, West Sussex: Fig. 113
The Warner Collection of Gulf States Paper Corporation, Tuscaloosa, Alabama: Fig. 114
Whitney Museum of American Art, New York: Fig. 115
Los Angeles County Museum of Art: Fig. 116
Hyland Granby Antiques, Dennis, Massachusetts: Fig. 117
Lowell Observatory, Flagstaff, Arizona: Fig. 118

INDEX